MW00679423

Chemical Engineering
License Problems and Solutions

Chemical Engineering
License Problems and Solutions

Dilip K. Das
Rajaram K. Prabhudesai

First Edition

Engineering Press Austin, TX

ISBN 1-57645-011-2
Library of Congress Cataloging in Publication Data
Das, Dilip K
Chemical Engineering License Problems and Solutions

Includes index
1. Chemical Engineering 1. Prabhudesai, Rajaram K.
II Title

TP 155.P75 1996 660.2 82-22890
ISBN 1-57645-011-2

Contents

Preface

Professional licensing and registration of all types of engineers including chemical engineers is being emphasized in all the states. Therefore, an increasing number of chemical engineers are taking the P.E. examination. The book *Chemical Engineering License Review* is primarily written to help chemical engineers taking the P.E. examination to quickly review and recapitulate the fundamentals of chemical engineering. It is also designed to help them to regain their confidence and facility in the application of chemical engineering fundamentals to solve a variety of problems within the time constraint and environment of the P.E. Exam. The example problems given in that book are general by design and not restricted to the type of problems set in the P.E. exam. This book, *Chemical Engineering License Problems and Solutions,* is an addition to the above book and is specifically written to familiarize the P.E. candidate to the types of problems that have been set in the past P.E. examinations and the type of problems that may be set in the near future examinations. It consists of only problems and their solutions. A thorough review of the material in these two books will enable the P.E. Candidate to prepare himself for the exam more adequately and face the actual exam with confidence and mental ease.

In addition to being helpful to P.E. candidates, this book will certainly be a good reference book for practicing chemical engineers and senior level chemical engineering students. The latter will especially benefit from the study of the problems and their solutions since it will enhance their understanding of the principles of chemical engineering through problem solving. Chemists and mechanical engineers working in the chemical industry also can use these books to familiarize themselves with the principles of chemical engineering in order to be more effective in their work.

Because the chemical engineers are expected to be conversant with engineering, SI and metric units, the problems in this book are designed to cover all of those units. At the end, a practice exam of 10 problems is given. The P.E. candidates should try to answer the questions in this exam as an exercise within the constraint of exam time before looking into the solutions given later in the book.

We have not added a separate nomenclature in this book also because of space limitations. Instead, we have used familiar notation. Whenever needed however, we have explained the same in the text.

Our sincere appreciation and thanks are due to Mrs. Vimal Prabhudesai and Mrs. Mala Das for their encouragement and forbearance during the completion of this book.

We wish good luck to our readers in their P.E. examinations and every success in their professional careers.

Dilip K. Das
Rajaram K. Prabhudesai

General Information and Suggestions to Candidates

Becoming A Professional Engineer

To achieve registration as a Professional Engineer there are four distinct steps: education, fundamentals of engineering (engineer-in-training) exam, professional experience, and finally, the professional engineer exam. These steps are described in the following sections.

Education

The obvious appropriate education is a B.S. degree in chemical engineering from an accredited college or university. This is not an absolute requirement. Alternative, but less acceptable, education is a B.S. degree in something other than chemical engineering, or from a non-accredited institution, or four years of education but no degree.

Fundamentals of Engineering (FE/EIT) Exam

Most people are required to take and pass this eight-hour multiple-choice examination. Different states call it by different names (Fundamentals of Engineering, E.I.T., or Intern Engineer) but the exam is the same in all states. It is prepared and graded by the National Council of Examiners for Engineering and Surveying (NCEES). Review materials for this exam are found in other books like Newnan: Engineer-In-Training License Review and Newnan and Larock: Engineer-In-Training Examination Review.

Experience

Typically one must have four years of acceptable experience before being permitted to take the Professional Engineer exam, but this requirement may vary from state to state. Both the length and character of the experience will be examined. It may, of course, take more than four years to acquire four years of acceptable experience.

Professional Engineer Exam

The second national exam is called Principles and Practice of Engineering by NCEES, but probably everyone else calls it the Professional Engineer or P.E. exam. All states, plus Guam, the District of Columbia, and Puerto Rico use the same NCEES exam.

Chemical Engineering Professional Engineer Exam Background

The reason for passing laws regulating the practice of chemical engineering is to protect the public from incompetent practitioners. Beginning about 1907 the individual states began passing title acts regulating who could call themselves a chemical engineer. As the laws were strengthened, the practice of certain aspects of chemical engineering was limited to those who were registered chemical engineers, or working under the supervision of a registered chemical engineer. There is no national registration law; registration is based on individual state laws and is administered by boards of registration in each of the states.

Examination Development

Initially the states wrote their own examinations, but beginning in 1966 the NCEES took over the task for some of the states. Now the NCEES exams are used by all states. This greatly eases the ability of a chemical engineer to move from one state to another and achieve registration in the new state. About 1600 chemical engineers take the exam each year. As a result about 23% of all chemical engineers are registered professional engineers.

The development of the chemical engineering exam is the responsibility of the NCEES Committee on Examinations for Professional Engineers. The committee is composed of people from industry, consulting, and education, plus consultants and subject matter experts. The starting point for the exam is a chemical engineering task analysis survey that NCEES does at roughly five to ten year intervals. People in industry, consulting and education are surveyed to determine what chemical engineers do and what knowledge is needed. From this NCEES develops what they call a "matrix of knowledge" that form the basis for the chemical engineering exam structure described in the next section.

The actual exam questions are prepared by the NCEES committee members, subject matter experts, and other volunteers. All people participating must hold professional registration. Using workshop meetings and correspondence by mail, the questions are written and circulated for review. The problems relate to current professional situations. They are structured to quickly orient one to the requirements, so the examinee can judge whether he or she can successfully solve it. While based on an understanding of engineering fundamentals, the problems require the application of practical professional judgement and insight. While four hours are allowed for four problems, probably any problem can be solved in 20 minutes by a specialist in the field. A professionally competent applicant can solve the problem in no more than 45 minutes. Multi-part questions are arranged so the solution of each succeeding part does not depend on the correct solution of a prior part. Each part will have a single answer that is reasonable.

Examination Structure

The ten problems in the morning four-hour session are regular computation "essay" problems. In the afternoon four-hour session all ten problems are multiple choice. In each category (Thermodynamics, Process design, Mass transfer, and so on) about half of the problems will be in the morning session and half in the afternoon. Engineering economics may appear as a component within one or two of the problems.

Note: The examination is developed with problems that will require a variety of approaches and methodologies including design, analysis, application, economic aspects, and operations.

Taking The Exam

Exam Dates

The National Council of Examiners for Engineering and Surveying (NCEES) prepares Chemical Engineering Professional Engineer exams for use on a Friday in April and October each year. Some state boards administer the exam twice a year in their state, while others offer the exam once a year. The scheduled exam dates are:

	April	October
1996		25
1997	18	31
1998	24	30
1999	23	29
2000	14	27

People seeking to take a particular exam must apply to their state board several months in advance.

Exam Procedure
Before the morning four-hour session begins, the proctors will pass out an exam booklet and solutions pamphlet to each examinee. There are likely to be civil, electrical, and mechanical engineers taking their own exams at the same time. You must solve four of the ten chemical engineering problems.

The solution pamphlet contains grid sheets on right-hand pages. Only work on these grid sheets will be graded. The left-hand pages are blank and are for scratch paper. The scratch work will not be considered in the scoring.

If you finish more than 30 minutes early, you may turn in the booklets and leave. In the last 30 minutes, however, you must remain to the end to insure a quiet environment for all those still working, and to insure an orderly collection of materials.

The afternoon session will begin following a one-hour lunch break. The afternoon exam booklet will be distributed along with an answer sheet. The booklet will have ten 10-part multiple choice questions. You must select and solve four of them. In an effort to make the exams more uniform, fewer problems will be presented on the exam. Our sample exam has five offered problems on the morning AM exam, four of which must be solved, and five offered problems for the afternoon PM exam of which four must be solved. An HB or #2 pencil is to be used to record your answers on the scoring sheet.

Exam-Taking Suggestions
People familiar with the psychology of exam-taking have several suggestions for people as they prepare to take an exam.

1. Exam taking is really two skills. One is the skill of illustrating knowledge that you know. The other is the skill of exam-taking. The first may be enhanced by a systematic review of the technical material. Exam-taking skills, on the other hand, may be improved by practice with similar problems presented in the exam format.

2. Since there is no deduction for guessing on the multiple choice problems, an answer should be given for all ten parts of the four selected problems. Even when one is going to guess, a logical approach is to attempt to first eliminate one or two of the five alternatives. If this can be done, the chance of selecting a correct answer obviously improves from 1 in 5 to, say, 1 in 3.

3. Plan ahead with a strategy. Which is your strongest area? Can you expect to see one or two problems in this area? What about your second strongest area? What will you do if you still must find problems in other areas?

4. Have a time plan. How much time are you going to allow yourself to initially go through the entire twelve problems and grade them in difficulty for you to solve them? Consider assigning a letter, like A, B, C and D, to each problem. If you allow 15 minutes for grading the problems, you might divide the remaining time into five parts of 45 minutes each. Thus 45 minutes would be scheduled for the first - and easiest - problem to be solved. Three additional 45 minute periods could be planned for the remaining three problems. Finally, the last 45 minutes would be in reserve. It could be used to switch to a substitute problem in case one of the selected problems proves too difficult. If that is unnecessary, the time can be used to check over the solutions of the four selected problems. A time plan is very important. It gives you the confidence of being in control, and at the same time keeps you from making the serious mistake of misallocation of time in the exam.

EXAMINATION SPECIFICATIONS
NCEES PRINCIPLES AND PRACTICE EXAMINATION
IN THE DISCIPLINE OF

CHEMICAL ENGINEERING

EFFECTIVE BEGINNING WITH APRIL 1995 EXAMINATION

		Number of Problems
A.	**FLUID MECHANICS** piping network problems; pump sizing or pump performance; compressor sizing or compressor performance; control valve selection problems; fluid flow through beds; two-phase flow; Bernoulli equation applications.	3
B.	**HEAT TRANSFER** industrial heat transfer including but not limited to the following: heat exchanger design and performance; energy conservation; conduction, especially insulation problems; convection; radiation, especially furnace design; evaporation.	3
C.	**CHEMICAL KINETICS** interpretation of experimental data and reaction rate modeling; commercial reactor design from rate model and/or product distribution; comparison of reactor types; reaction control.	2
D.	**MASS AND ENERGY BALANCES** process stoichiometry and material balances; process energy balances; conservation laws.	4
E.	**MASS TRANSFER** typical applications including but not limited to the following: gas absorption and stripping; distillation; liquid-liquid extraction and leaching; humidification and dehumidification; drying.	3
F.	**PLANT DESIGN** process and equipment design including but not limited to the following: optimization of design; general safety considerations; environmental and waste treating; solids separation; vapor-liquid separations; flow sheets; HAZOP (hazard and operational) analysis; fault tree analysis; scheduling techniques; sizing and fabrication of equipment; material selection; life cycle cost; physical and chemical properties of matter; strength of materials; crystallographic structure; phase diagrams (metallurgical); latent heat; PVT data and relationships; molecular structure; sensors; transmitters and controllers; control loops; simulation.	3
G.	**THERMODYNAMICS** estimation and correlation of physical properties; chemical equilibrium; heats of reaction; application of first and second laws; vapor-liquid equilibrium; combustion; refrigeration.	2
	Total number of problems =	20

Note: The examination is developed with problems that will require a variety of approaches and methodologies including design, analysis, application, and operations. Some problems may include engineering economics analyses.

5. Read all five multiple choice answers before making a selection. The first answer in a multiple choice question is sometimes a plausible decoy - not the best answer.

6. Do not change an answer unless you are absolutely certain you have made a mistake. Your first reaction is likely to be correct.

7. Do not sit next to a friend, a window, or other potential distractions.

Exam Day Preparations

There is no doubt that the exam will be a stressful and tiring day. This will be no day to have unpleasant surprises. For this reason we suggest that an advance visit be made to the examination site. Try to determine such items as:

1. How much time should I allow for travel to the exam on that day? Plan to arrive about 15 minutes early. That way you will have ample time, but not too much time. Arriving too early, and mingling with others who also are anxious, will increase your anxiety and nervousness.

2. Where will I park?

3. How does the exam site look? Will I have ample work space? Where will I stack my reference materials? Will it be overly bright (sunglasses) or cold (sweater), or noisy (earplugs)? Would a cushion make the chair more comfortable?

4. Where is the drinking fountain, lavatory facilities, pay phone?

5. What about food? Should I take something along for energy in the exam? A bag lunch during the break probably makes sense.

What To Take To The Exam

The NCEES guidelines say you may bring the following reference materials and aids into the examination room for your personal use only:
1. Handbooks and textbooks
2. Bound reference materials, provided the materials are and remain bound during the entire examination. The NCEES defines "bound" as books or materials fastened securely in its cover by fasteners which penetrate all papers. Examples are ring binders, spiral binders and notebooks, plastic snap binders, brads, screw posts, and so on.
3. Battery operated, silent non-printing calculators.

At one time NCEES had a rule that did not permit "review publications directed principally toward sample questions and their solutions" in the exam room. This set the stage for restricting some kinds of publications from the exam. State boards may adopt the NCEES guidelines, or adopt either more or less restrictive rules. Thus an important step in preparing for the exam is to know what will - and will not - be permitted. We suggest that if possible you obtain a written copy of your state's policy for the specific exam you will be taking. Recently there has been considerable confusion at individual examination sites, so a copy of the exact applicable policy will not only allow you to carefully and correctly prepare your materials, but also will insure that the exam proctors will allow all proper materials.

As a general rule we recommend that you plan well in advance what books and materials you want to take to the exam. Then they should be obtained promptly so you use the same materials in your review that you will have in the exam.

License Review Books
There are two rules that we suggest you follow in selecting license review books to insure that you obtain up-to-date materials:

1. Consider the purchase only of materials that have a 1993 or more recent copyright. The exam used to be 20 essay questions, including a full scale engineering economics problem. The engineering economics problem is gone (at least as a separate question) and half the remaining 10 problems are now multiple choice.

2. Even if a license review book has a recent copyright date, is the content up to date? Books with older content probably will also lack the afternoon multiple choice problems.

Textbooks
If you still have your university textbooks, we think they are the ones you should use in the exam, unless they are too out of date. To a great extent the books will be like old friends with familiar notation.

Bound Reference Materials
The NCEES guidelines suggest that you can take any reference materials you wish, so long as you prepare them properly. You could, for example, prepare several volumes of bound reference materials with each volume intended to cover a particular category of problem. Use tabs so specific material can be located quickly. If you do a careful and systematic review of chemical engineering, and prepare a lot of well organized materials, you just may find that you are so well prepared that you will not have left anything of value at home.

Other Items
Calculator - NCEES says you may bring a battery operated, silent, non-printing calculator. You need to determine whether or not your state permits pre-programmed calculators. Extra batteries for your calculator are essential, and many people feel that a second calculator is also a very good idea.

Clock - You must have a time plan and a clock or wristwatch.

Pencils - You should consider mechanical pencils that you twist to advance the lead. This is no place to go running around to sharpen a pencil, and you surely do not want to drag along a pencil sharpener.

Eraser - Try a couple to decide what to bring along. You must be able to change answers on the multiple choice answer sheet, and that means a good eraser. Similarly you will want to make corrections in the essay problem calculations.

Exam Assignment Paperwork - Take along the letter assigning you to the exam at the specified location. To prove you are the correct person, also bring something with your name and picture.

Items Suggested By Advance Visit - If you visit the exam site you probably will discover an item or two that you need to add to your list.

Clothes - Plan to wear comfortable clothes. You probably will do better if you are slightly cool.
Box For Everything - You need to be able to carry all your materials to the exam and have them conveniently organized at your side. Probably a cardboard box is the answer.

Exam Scoring

Essay Questions

The exam booklets are returned to Clemson, SC. There the four essay question solutions are removed from the morning workbook. Each problem is sent to one of many scorers throughout the country.

For each question an item specific scoring plan is created with six possible scores: 0, 2, 4, 6, 8, and 10 points. For each score the scoring plan defines the level of knowledge exhibited by the applicant. An applicant who is minimally qualified in the topic is assigned a score of 6 points. The scoring plan shows exactly what is required to achieve the 6 point score. Similar detailed scoring criteria are developed for the two levels of superior performance (8 and 10 points) and the three levels of inferior performance (0, 2, and 4 points). Every essay problem submitted for grading receives one of these six scores. The scoring criteria may be based on positive factors, like identifying the correct computation approach, or negative factors, like improper assumptions or calculation errors, or a mixture of both positive and negative factors. After scoring, the graded materials are returned to NCEES, which reassembles the applicants work and tabulates the scores.

Multiple Choice Questions

Each of the four multiple choice problems is 10 points, with each of the ten questions of the problem worth one point. The questions are machine scored by scanning. The input data are evaluated by computer programs to do error checking. Marking two answers to a question, for example, would be detected and no credit given. In addition, the programs identify those questions with statistically unlikely results. There is, of course, a possibility that one or more of the questions is in some way faulty. In that case a decision will be made by subject matter experts on how the situation should be handled.

Passing The Exam

In the exam you must answer eight problems, each worth 10 points, for a total raw score of 80 points. Since the minimally qualified applicant is assumed to average six points per problem, a raw score of 48 points is set equal to a converted passing score of 70. Stated bluntly, you must get 48 of the 80 possible points to pass. The converted scores are reported to the individual state boards in about two months, along with the recommended pass or fail status of each applicant. The state board is the final authority of whether an applicant has passed or failed the exam.

Although there is some variation from exam to exam, the following gives the approximate passing rates:

Applicant's Degree	Percent Passing Exam
Engineering from accredited school	62%
Engineering from non-accredited school	50
Engineering Technology from accredited school	42
Engineering Technology from non-accredited school	33
Non-Graduates	36
All Applicants	56

Although you want to pass the exam on your first attempt, you should recognize that if necessary you can always apply and take it again.

This Book

This book is organized to cover the chemical engineering professional engineer (principles and practice) exam.

NCEES does not allow their problems to be reproduced, so none of the problems in this book came from them. Each one is structured to approximate the scope and difficulty of the actual exam problems you will encounter. The National Council of Examiners for Engineering and Surveying (NCEES), which prepares the chemical engineering examination, calls it an open book examination. Most states accept this and allow applicants to bring textbooks, handbooks and any bound reference materials to the exam. A few states, however, do not permit review publications directed principally toward sample problems and their solutions.

MATERIAL AND ENERGY BALANCES

1-1: A manufacturer of high grade lime uses very high quality $CaCO_3$ which analyses 98% $CaCO_3$, 1.2% $MgCO_3$ and 0.8 % $Al_2O_3.SiO_2$. The $CaCO_3$ (limestone) is fed to a rotary kiln after heating it to 400 ^{0}F in a preheater. The calcined product leaves the rotary kiln at 1832 ^{0}F. Assume 100 % conversion of $CaCO_3$. Given the following data, calculate

(a) The theoretical heat required to be supplied to the kiln if the kiln feed rate is 15 tons per hour.

(b) Oil firing rate if the overall thermal efficiency of kiln operation is 80 %.

component	MW	$(\Delta H_f)_{25}$ kcal/gmol	C_P - cal/gmol (T is in K)
$CaCO_3$	100.0	- 289.5	$C_P = 19.68 + 0.01189T$
$MgCO_3$	84.0	- 261.7	$C_P = 16.9$
$Al_2O_3.SiO_2$	162.0	- 648.7	$C_P = 40.79 + 0.004763T$
CaO	56.0	- 151.7	$C_P = 10.0 + 0.00484T$
MgO	40.0	- 43.84	$C_P = 10.6 + 0.001197T$
CO_2	44.0	- 94.02	$C_P = 10.34 + 0.00274T$

Fuel oil composition: C = 86 % , H = 12% , N = 2% ; HHV of fuel oil = 19200 Btu/lb

1-2: A vessel contains 150 ft^3 of a solution containing 20 % of NaCl. Another solution containing 1% NaCl is run into the vessel at a rate of 3 ft^3/min. At the same time, solution from the tank is run out at a rate of 3 ft^3/min. How long will it take to reduce the concentration of NaCl in solution in the tank to 5 %. Assume tank contents are perfectly mixed by agitation at any instant and the densities can be assumed equal and constant.

1-3: The ultimate analysis of the coal fed to a boiler to generate steam is as follows

Carbon =76.0% Oxygen =3.2% Sulfur = 1.6%
Hydrogen = 6 % Nitrogen = 1.4 %
Ash=11.8%

The coal also contains 4.6 lb free moisture/100 lb dry coal. The temperature of coal is 70 ^{0}F. The dry refuse from the ashpit contains 17% unburned carbon. Its temperature is 275 ^{0}F and its specific heat is 0.23 Btu/lb.^{0}F. The Orsat analysis of the flue gases gives the following composition:

$CO_2 + SO_2 = 12$ % $O_2 = 5.9$ % CO = 1.2 %

The air is supplied to the burner at 70 ^{0}F dry bulb and 60 ^{0}F wet bulb. Pressure = 755 mm Hg
Calculate

1. Weight of refuse per ton of coal
2. Weight of dry gaseous products per 100 lb of coal.
3. Weight of dry air supplied per 100 lb coal fired.
4. Weight of moisture in flue gas per 100 lb coal fired

5. Total flue gas per 100 lb coal fired
6. Excess air used per 100 lb coal fired

Solutions

1-1:

Material balance : Basis 100 lb of feed (limestone) to the rotary kiln.

Component	Feed	Products		Product CO_2
	lb		lb	lb
$CaCO_3$	98.00	CaO	54.88	43.12
$MgCO_3$	1.20	MgO	0.57	0.63
Al_2O_3 SiO_2	0.8	$Al_2O_3SiO_2$	0.80	-
	100.00		56.25	43.75

Since reaction data are available at 25 °C, it is convenient to choose the following path for calculations

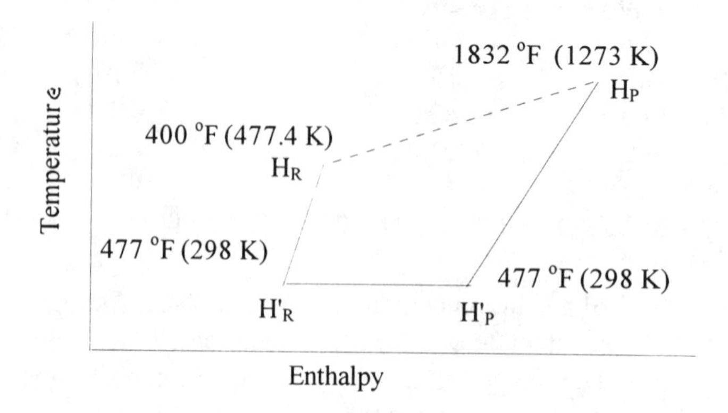

$$Q = \Sigma H_P + \Sigma \Delta H_{25} - \Sigma H_R$$

$T_{ref} = 273 + 25 = 298$ K T_i =400 °F = 204.4 C = 273 + 204.4 = 477.4 K

T_o = 1832 °F = 1000 °C = 1273 K

Heats of reaction :

For $CaCO_3$ decomposition, $\Delta H_R = (-94.02 - 151.7) - (-289.5) = 43.78$ kcal/gmole
$= (43.78 \times 1800)/100 = 788.04$ Btu/lb $CaCO_3$

For $MgCO_3$ decomposition, $\Delta H_R = (-94.02 - 143.84) - (-261.7) = 23.84$ kcal/gmole

$= (23.84 \times 1800)/84 = 510.86$ Btu/lb $MgCO_3$

$\Sigma \Delta H_{25} = 98(788.04) + 1.2(510.86) = 77841$ Btu/100 lb of limestone feed

(In the following calculations , the factor 10^{-3} is used to convert cal into kcal.)

Enthalpy changes for reactants from 400 °F to 77 °F ($\Delta H_{reactants}$ = $H_R' - H_R$)

$CaCO_3$ $H_R = 10^{-3}[19.68(477.4-298)+0.005945(477.4^2-298^2]$
$= 4.358$ kcal/gmole $= 78.44$ Btu/lb $CaCO_3$

$MgCO_3$ $H_R = 10^{-3}[16.9(477.4-298)] = 3.032$ kcal/gmole $= 64.97$ Btu/lb

$Al_2O_3SiO_2$ $H_R = 10^{-3}[40.79(477.4-298)+0.0023815(477.4^2-298^2]$
$= 7.65$ kcal/gmole $= 85$ Btu/lb $Al_2O_3SiO_2$

Enthalpy changes for products from 77 °F to 1832 °F (ΔH Products $= H_P - H_P'$)

CaO $H = 10^{-3}[10.0(1273-298)+0.00242(1273^2-298^2)]$
$= 13.457$ kcal/gmole $= 432.6$ Btu/lb CaO

MgO $H = 10^{-3}[10.86(1273-298)+0.0005985(1273^2-298^2)]$
$= 11.505$ kcal/gmole $= 517.3$ Btu/lb MgO

$Al_2O_3SiO_2$ $H = 10^{-3}[40.79(1273-298)+0.0023815(1273^2-298^2)]$
$= 43.418$ kcal/gmole $= 482.4$ Btu/lb $Al_2O_3SiO_2$

CO_2 $H = 10^{-3}[10.34(1273-298)+0.00137(1273^2-298^2)]$
$= 12.18$ kcal/gmole $= 498.3$ Btu/lb CO_2

Q = Total enthalpy change $= \Sigma H_P + \Sigma(\Delta H_R) - \Sigma H$
$= \{54.88(432.6) + 0.57(517.73) + 0.8(482.4) + 43.75(498.3)\} + 77841 - \{98(78.44) + 1.2(64.97) + 0.8(85)\}$

$= 116231$ Btu/100 lb limestone feed

(a) Kiln feed rate is 15 tons per hour = 30000 lb/h
Theoretical heat to be supplied per hour $= \frac{30000}{100} \times 116231 = 34.9 \times 10^6$ Btu/h
With 80 % thermal efficiency, heat to be supplied = $34.9\text{x}10^6/0.8 = 43.6\text{x}10^6$ Btu/h

(b) Oil firing rate
HHV of fuel oil = 19200 Btu/lb
Hydrogen in fuel = 0.12 lb per lb of oil
lb-moles of hydrogen = 0.12/2 = 0.06 lb-mole per lb of oil
water produced = 0.06x18 = 1.08 lb per lb of oil
λ- latent of vaporization of water at 77 °F = 1050.1 Btu/lb of water
Therefore, LHV = net heating value = 19200 - 1050.1x1.08 = 18066 btu/lb oil
Then oil firing rate = $43.6\text{x}10^6$ / 18066 = 2413 lb/h

1-2:

(This problem illustrates the method of solving unsteady state problems.)
First take overall material balance over an incremental time interval $\Delta\theta$
Let V = volume, F_i = vol. flow rate into the tank , F_0 = vol. flow rate out of tank.
Overall mass balance gives

$$\rho V_E - \rho V_B = \rho F_i \Delta\theta - \rho F_o \Delta\theta$$

where V_E = Volume of solution in tank at end of time interval $\Delta\theta$

V_B = Volume of solution in tank at beginning of time interval $\Delta\theta$
ρ = density of solution assumed constant and equal for all streams

In a difference form, the mass balance equation after cancelling out ρ becomes

$$\frac{\Delta V}{\Delta\theta} = F_i - F_o$$

Taking limit as $\Delta\theta \to 0$ gives the differential equation

$$\frac{dV}{d\theta} = F_i - F_o$$

However since inlet and outlet flow rates are given equal, the above equation reduces to

$$\frac{dV}{d\theta} = 0$$

Now take a material balance on solute over a time interval $\Delta\theta$

$$\rho V_E C_E - \rho V_B C_B = \rho F_i C_i \Delta\theta - \rho F_o \bar{C} \Delta\theta$$

Where C_E = Concentration of solute at the end of the accounting interval $\Delta\theta$
C_B = concentration of the solute at the beginning of the accounting period $\Delta\theta$
$\bar{C}$ = average concentration of solute over accounting period $\Delta\theta$
The difference form can be written after canceling out ρ as

$$\frac{\Delta(VC)}{\Delta\theta} = F_i C_i - F_o \bar{C}$$

taking limit as $\Delta\theta \to 0$, $\frac{d(VC)}{d\theta} = F_i C_i - F_o C$

where C is instantaneous concentration of the solute at time θ

The equation can be expanded as $V\frac{dC}{d\theta} + C\frac{dV}{d\theta} = F_i Ci - F_0 C$

Since $\frac{dV}{d\theta} = 0$, the above reduces to $V\frac{dC}{d\theta} = F_i C_i - F_o C$

substitution of the given values for the flow rates yields the following equation

$$150\frac{dC}{d\theta} = 3(0.01) - 3C$$

separating variables, one gets

$$150\,\frac{dC}{0.03 - 3C} = d\theta$$

or

$$50\int_{0.2}^{0.05} \frac{dC}{0.01 - C} = \int_0^{\theta} d\theta$$

With initial condition, when $\theta = 0$, C = C_B = 0.2

Integration of the equation gives

$$-50 \ln (0.01 - C) = \theta + K$$

where K = Integration constant.

Applying the initial condition , the constant of integration is obtained as

$$-50 \ln(0.01 - C_B) = K$$

The solution of the equation becomes

$$-50 \ln (0.01 - C) = \theta - 50 \ln (0.01 - C_B)$$

Simplification gives the solution $C = 0.19e^{-0.02\theta} + 0.01$

To reduce the concentration to 5% , C = 0.05. putting this value in the equation gives

$$0.05 = 0.19e^{-0.02} + 0.01 \quad \text{or} \quad e^{-0.02\theta} = \frac{0.05 - 0.01}{0.19} = 0.21052$$

Taking logarithms , $-0.02\theta \ln e = \ln(0.210526)$ or $-0.02\,\theta = -1.5581$

which gives $\theta = (-1.5581/0.02) =$ **77.9 Minutes**.

Alternatively, the following shorter solution is also possible.

Integrating with limits, (a) $\theta = 0$, C = 0.2, and (b) $\theta = \theta$, C = 0.5 gives

$$\theta = -50[\ln(C - 0.01)]_{0.2}^{0.05}$$

$$= -50[\ln(0.05 - 0.01) - (\ln(0.2 - 0.01))]$$

$$= -50(-1.5581) = \mathbf{77.9\ minutes}$$

1-3:

(1) Weight of refuse per ton of coal

weight of ash in coal 11.8 lb per 100 lb of coal.

carbon in refuse = (0.17/0.83) 11.8 = 2.42 lb

total refuse = 2.42 + 11.8 = 14.22 lb /100 lb coal

Refuse per ton of coal = (2000/100) 14.22 = 284.4 lb = $\frac{284.4}{2000}$ = 0.1422 ton /ton of coal.

(2) Weight of dry gaseous products per 100 lb of coal

carbon in coal = 76 lb = 6.34 lb-atom

carbon in refuse = 2.42 lb = 0.2 lb-atom

carbon in stack gases = 76 - 2.42 = 73.58 = 6.132 lb-atom

sulfur in stack gas as SO_2 = 1.6 lb = 1.6/32 = 0.05 lb-atom.
SO_2 in stack gas = 0.05 lb-mol
Let x moles of CO_2 in dry gas from coal and y moles of CO in dry gas

$$\frac{x + 0.05}{y} = \frac{12}{1.2} = 1 \qquad \text{and} \qquad x + y = 6.132$$

Solving these two equations y = 0.562 and x = 5.57 lb-mol
Nitrogen from coal = 1.4 lb = 1.4/28 = 0.05 lb-mol

lb-moles of dry gas from coal:

Carbon in stack gas = 6.132 lb-atom
Moles of SO_2 + CO_2 = 0.05 + 5.57 = 5.62
Percentage of SO_2 = (0.05/5.62) x 12= 0.107 %
Percentage of CO_2 = 12 - 0.107 = 11.893 %
Percentage of CO_2 + CO in stack gas = 11.893 + 1.2 = 13.093 %
Then total dry gas = (5.57 + 0.562)/0.13093 = 46.834 lb-mol
O_2 in dry gaseous product = 0.059 x 46.834 = 2.763 lb-mol

Total dry gaseous products .
Basis : 100 lb of coal fired.

component	lb-mol	MW	lb
CO_2	5.57	44	245.08
CO	0.562	28	15.74
SO_2	0.05	64	3.2
N_2 (by difference)	37.889	28	1060.89
O_2	2.763	32	88.416
	46.834		1413.326

Average molecular weight = 1413.326/46.834 = 30.18

(3) Weight of dry air supplied :

Nitrogen from dry air = 37.889 - 0.05 = 37.839 lb-mol per 100 lb coal fired
Dry air supplied = 37.839/0.79 = 47.9 lb-mol = 47.9 x 29 = 1389 lb

(4) Weight of moisture in flue gas per 100 lb coal fired

Air humidity at dry-bulb 70 ^oF and 60 ^oF temperature = 0.0089 lb/lb of dry air
= 0.0089/0.62 = 0.0144 lb-mol/lb-mol of dry air
Moisture from air = 47.9 x 0.0144 = 0.69 lb-mol per 100 lb coal
Free moisture from coal = 4.6 lb = 4.6/18 = 0.26 lb-mol per 100 lb of coal
Water vapor from combustion of hydrogen in coal = 6/2 = 3 lb-mol
(1 mol H_2 1 lb-mol of water, mols of H_2 in coal = 6/2 = 3 lb-mol/100 lb coal)
Weight of moisture in flue gas = 18(0.69 + 0.26 + 3.0) = 71.1 lb per 100 lb coal.

(5) Total flue gas

Basis 100 lb coal fired

Total flue gas = dry gas + moisture = 46.834 + 3.95 = 50.784 lb-mol

Flue gas temperature exiting from boiler = 480 ^{0}F

Assume negligible pressure drop for flue gas stream in the boiler

Then volume of flue gas = $50.784 \times 359 \times \frac{760}{755} \times \frac{940}{492} = 35063$ ft^3

(6) Excess air used

Theoretical amount of oxygen required: Basis 100 lb coal fired.

Component	lb	lb-mol	O_2 required
Carbon	76	6.34	6.34
Hydrogen	6	3.0	1.5
Sulfur	1.6	0.05	0.05
Oxygen	3.2	0.10	- 0.10
			7.79

Net oxygen required = 7.79 lb-mol

Oxygen in flue gas due to incomplete combustion of CO to CO_2 = 0.562/2 = 0.281 lb.mol

Oxygen in flue gas due to non-combustion of C in refuse = 2.42/12 = 0.2017 lb.mol

Therefore , true excess oxygen in the flue gas = 2.763 - 0.281 - 0.2017 = 2.2803 lb.mol

% excess dry air = (2.2803/7.79) x 100 = **29.27 %**

{ Alternatively, dry air supplied = 47.9 lb.mol

Theoretical air required = 7.79/0.21 = 37.1 lb.mol

Therefore, % excess dry air = [(47.9 - 37.1)/37.1] x100 = 29.11 %}

Fluid Mechanics

2-1 The following data of a pumping system is available:

Pump capacity: 250 gpm
Liquid viscosity: 100 cP
Liquid specific gravity: 1.2
Liquid vapor pressure: 2.89 psia
Roughness factor of pipe: .00015 ft

	SUCTION SIDE	DISCHARGE SIDE
Static head(ft)	3	4
Operating pressure(psia)	14.7	14.7
Pipe inside diameter(inch)	5.629	4.026
Straight length of pipe(ft)	26	162
Number of sudden contractions	1	3
Number of ball valves	2	2
Number of check valves (swing,tilting seat)	0	2
Number of globe valves(bevel seat)	0	1
Number of 90-degree elbows(standard, flanged)	6	27
Number of tees(flow through run, welded)	1	1
Number of tees(flow-out through branch, welded)	2	6
Number of sudden expansions	0	2
Number of orifices(beta-ratio=0.5)	0	2
Control valve pressure drop(psi)	0	400
Exchanger pressure drop(psi)	0	30
Other equipment pressure drop(psi)	2	75

If any necessary information is missing, please document assumptions.
Answer the following multiple-choice questions:

2-1.1 What is the static suction head(ft) of the pump?
(A) 4 (B) 3 (C) 1 (C) -1 (D) 0

2-1.2 What is the absolute pressure on the liquid surface of the suction vessel in feet of liquid?
(A) 33.96 (B) 17.64 (C)28.30 (D) 14.7

2-1.3 The absolute vapor pressure of the liquid
at the pumping temperature in feet of liquid is approximately :
(A) 7 (B) 6 (C) 8 (D) 3

2-1.4 The frictional loss in the suction pipe in feet of liquid is approximately:
(A) 12 (B) 2 (C) 6 (D) Cannot be found with the information given

2-1.5 If the frictional loss in feet of liquid in the suction pipe is 6.17, and the linear velocity(ft/s) through the suction pipe is 3.2234, then $NPSH_A$ is close to:
(A) 5 (B) 10 (C) 20 (D) -20

2-1.6 The frictional pressure drop(psi) in the discharge pipe due to the fittings and straight pipe excluding the orifice, control valve and other equipment is conservatively:
(A) 38 (B) 6 (C) 24 (D) 8

2-1.7 The frictional pressure drop through two orifices in psi, without considering pressure recovery is close to:
(A) 1 (B) 2 (C) 5 (D) 25

2-1.8 The total frictional pressure drop in psi through control valve, exchanger and other equipment in the discharge side is:
(A) 505 (B) 400 (C) 25 (D) 75

2-1.9 If the total frictional pressure drop(psi) in suction line is 11.88, and total frictional pressure drop in discharge line is 567.69 psi, the the TDH(ft) is approximately:
(A) 1117 (B) 500 (C) 250 (D) 600

2-1.10 If the TDH(ft) is 1116.67 then the hydraulic horsepower is
(A) 20 (B) 10 (C) 30 (D) 84.6

2-2. One hundred gallons per minute of water flows through two junctions connected by two parallel lines. Calculate the flow through each parallel line, and the pressure drop between the junctions. Data available are:
Equivalent length of the first line: 100 ft
Inside diameter of the first line: 0.824 inch
Equivalent length of the second line: 200 ft
Inside diameter of the second line: 1.610 inch
Viscosity of water: 1 cP
Density of water: 62.4 lb/ft^3
Fanning friction factor, $f = 0.054/(N_{Re})^{0.2}$ for turbulent flow

2-3. Nitrogen is supplied through a regulator valve with port size equal to the internal diameter of the line which is 1" schedule 40 (inside diameter=1.049") to a distillation column operating at 5 psig. If the pressure and temperature of nitrogen are 100 psig and 90 °F respectively before the regulator, calculate the flow of nitrogen if the regulator fails to full open position and the total equivalent length of pipe is 25 feet. Assume roughness factor of pipe,ϵ, is 0.00015 ft and Fanning friction factor, f, for fully turbulent flow is given by:

$$f = 0.0625[\log_{10}(3.7D/\epsilon)]^{-2}$$

where D is internal diameter of pipe in ft. The ratio of specific heat of gas at constant pressure to that at constant volume is 1.4.

2-4. A vertical, cylindrical tank, dished top and flat bottom, 10 ft in internal diameter and 20 ft high (bottom to upper tangent line), open to atmosphere, is filled with water upto the upper tangent line. A 1.049" diameter leak develops on the side wall with the center of the leak being 1 ft vertically above the bottom of the tank.

2-4.1 How long will it take for the level of the tank to drop to 5 ft above the bottom?

2-4.2 How long will it take for the same drop in level if the leak is fitted with a horizontal pipe with an internal diameter of 1.049" and a straight length of 10 ft?

SOLUTIONS

2-1

2-1.1

Z_1 = Static suction head = 3 ft, given. **Answer: B**

2-1.2

Z_a = Absolute pressure on the liquid surface of suction vessel= 14.7 psia, given.
= 14.7 x 2.31/s: (Chem. Eng. License Review, equation 2-33)
= 14.7 x 2.31/1.2 = 28.30 ft **Answer: C**

2-1.3

Z_v = Absolute vapor pressure of the liquid at pumping temperature
= 2.89 psia, given.= 2.89 x 2.31/1.2 = 5.56 ft **Answer: B**

2-1.4

Calculation of Z_f, the frictional loss in suction piping:

Cross section of pipe = $\pi x D^2/4 = \pi x(5.629/12)^2/4 = 0.17281$ ft^2.
u = linear velocity, ft/s = (250/7.48)/(60x0.17281) = 3.2234 ft/s
$N_{Re} = Du\rho/\mu$ = 5.629x3.2234x62.4x1.2/(12x100x0.000672) = 1684.8482
Flow is laminar.
Therefore, Fanning friction factor, f = 16/N_{Re} = 16/1684.8482 = 0.0094964
Calculate K values of fittings in the suction line(please refer to Table 2.1 of CELR):

fittings at pump suction	K values
Pipe entrance, sharp edged	0.50
2 ball valves: 2x4fL_e/D = 2x4x.0094964x3 =	0.23
6 90° elbows: 6x4fL_e/D = 6x4x.0094964x30 =	6.84
1 tee(flow through run): 4fL_e/D = 4x.0094964x8 =	0.30
2 tees(flow through branch):2x4fL_e/D = 2x4x.0094964x58 =	4.41
	ΣK = 12.28

2-4

$Z_f = (\Sigma K + 4fL/D)(u^2/2g_c)$ + other equipment losses
$=(12.28+4x.0094964x26x12/5.629)x(3.2234)^2/(2x32.2) + 2x2.31/1.2 = 6.17$ ft
Answer: C

2-1.5

The available NPSH is given by:
$NPSH_A = Z_1 + Z_a - Z_v - Z_f + u^2/2g_c$, ft of liquid: (CELR equation 2-38)

$NPSH_A = 3 + 28.30 - 5.56 - 6.17 + (3.2234)^2/64.4) = 19.73$ ft **Answer: C**

2-1.6

Computation of pressure drop due to fittings in discharge side:

Cross section area of pipe = $\pi x D^2/4 = \pi x(4.026/12)^2/4 = 0.088404$ ft^2.
u = linear velocity,ft/s=(250/7.48)/(60x0.088404) = 6.3011 ft/s.
$N_{Re} = Du\rho/\mu$=4.026x6.3011x62.4x1.2/(12x100x0.000672) = 2355.6212
Flow is transitional.
ϵ/D = ratio of roughness factor to diameter =0.00015x12/4.026 = 0.00045.
From figure 2.7 of CELR, the Fanning friction factor, f = 0.013(conservative).

fittings in discharge side	K-values
3 pipe entrances, sharp edged= 3x0.5 =	1.50
2 ball valves: $2x4fL_e/D$ = 2x4x.013x3 =	0.31
2 check valves: $2x4fL_e/D$=2x4x.013x100=	10.40
1 globe valve: $1x4fL_e/D$=1x4x.013x340 =	17.68
27 90° elbows: $27x4fL_e/D$=27x4x.013x30 =	42.12
1 tee(run, welded):$1x4fL_e/D$=1x4x.013x8=	0.42
6 tees(branch, welded):	
$6x4fL_e/D$ = 6x4x.013x58 =	18.10
2 sudden expansions: 2x1 =	2.00
ΣK =	92.53

Pressure drop in fittings and straight line = $(\Sigma K + 4fL_e/D)(u^2/2g_c)$
= (92.53 + 4x.013x162x12/4.026)(6.3011)^2/64.4 = 72.53 ft
= 72.53x.433xs = 72.53x.433x1.2 = 37.69 psi
Answer: A

2-1.7

Pressure drop in the orifice may be computed from equation (2-18) of CELR:

$$W = \rho A_0 C_0 Y[(2g\Delta H)/(1-\beta^4)]^{0.5}$$

Where:
ρ = fluid density = 1.2x 62.4 = 74.88 lb/ft^3
W = flow rate of fluid = 250gpm = 250x74.88/(7.48x60) = 41.7112 lb/s

A_0 = orifice area = $\pi(D_o)^2/4 = \pi(0.5x4.026/12)^2/4 = 0.0221$ ft^2
C_0 = orifice coefficient = 0.62, assumed for turbulent flow.
Y = expansion factor = 1, assumed for incompressible fluid.
g = 32.2 ft/s^2.
β = ratio of orifice dimeter to pipe diameter = 0.5
Substituting in above equation, the pressure drop through one orifice is obtained by:
ΔH = 24.06 ft
Pressure drop through two orifices = 48.12 ft = 48.12x.433x1.2 = 25 psi.
Some of this pressure drop is recovered downstream of the meters, but is neglected for a conservative estimate. **Answer: D**

2-1.8

Pressure drop through control valve, exchanger and other equipment in the discharge side
= (400 + 30 + 75) = 505 psi **Answer: A**

2-1.9

Total dynamic head

From equation (2-37) of CELR, the total dynamic head(TDH) is given by:

$$TDH(ft) = (P_2 - P_1 + \Delta P_{f1} + \Delta P_{f2})2.31/s + Z_2 - Z_1$$

P_2 = Pressure in discharge vessel, psia = 14.7
P_1 = Pressure in suction vessel, psia = 14.7
ΔP_{f1} = Pressure drop in suction line, psi = 11.88 psi
ΔP_{f2} = Pressure drop in discharge line, psi. This includes pressure drop in fittings, line, orifice, cotrol valve, exchanger and other equipment = 567.69 psi
Z_2 = discharge side static head = 4 ft
Z_1 = suction side static head = 3 ft

TDH(ft) = (14.7 - 14.7 + 11.88 + 567.69)2.31/1.2 + 4 - 3
= 1116.67 ft **Answer: A**

2-1.10

The hydraulic horsepower
= Q(TDH)s/3960 Equation (2-59), CELR
= 250x1116.67x1.2/3960 = 84.6 **Answer: D**

2-2.

From equation (2-26), CELR:

$$\Delta P_f = (2fu^2L\rho/(g_cD) \qquad \text{equation (1)}$$

Given by the problem:

$$f = 0.054/(N_{Re})^{0.2} \qquad \text{equation (2)}$$

2-6

By definition:
$N_{Re} = Du\rho/\mu$ equation (3)

The linear velocity u(ft/s), flow rate Q(gpm), and diameter D(ft) are related by:
$u = 4Q/(60x7.48x\pi xD^2)$ equation (4)

Combining equations 1,2, 3, and 4:
$\Delta P_f = 8.7233x10^{-8}\mu^{0.2}\rho^{0.8}Q^{1.8}L/D^{4.8}$ equation (5)

Since for parallel flow paths ΔP is same for each path, the flow rate for each path may be derived from equation 5 as:
$Q = KD^{2.67}/L^{0.56}$ equation (6)

Where K is constant for flow through parallel branch with constant density and viscosity of the fluid.

If Q_1 = flow through 100 ft branch with the internal diameter of D_1(0.824 in) and Q_2 = flow through 200 ft branch of internal diameter, D_2(1.61 in), then:

$$\frac{Q1}{Q1 + Q2} = \frac{\dfrac{KD_1^{2.67}}{L_1^{0.56}}}{\dfrac{KD_1^{2.67}}{L_1^{0.56}} + \dfrac{KD_2^{2.67}}{L_2^{0.56}}}$$

$$= \frac{1}{1+\left(\dfrac{D_2}{D_1}\right)^{2.67} \times \left(\dfrac{L_1}{L_2}\right)^{0.56}}$$

$$= \frac{1}{1+\left(\dfrac{1.61}{0.824}\right)^{2.67} \times (0.5)^{0.56}}$$

$$= 0.1978$$

Therefore, flow through 100 ft branch = .1978x100 = 19.78 gpm, and flow through 200 ft branch, by difference, is 80.22 gpm.

The pressure drop between the junctions may be computed by considering flow through any of the two parallel lines. Consider flow through the 200 ft branch.
From equation 5:

$$\Delta P_f = \frac{8.7233\times 10-8\times(0.000672)0.2\times(62.4)0.8\times(80.22)1.8\times 200}{\left(\dfrac{1.61}{12}\right)^{4.8}}$$

$$= 4553.5 \text{ lbf / ft2} = 31.62 \text{ psi}$$

Alternatively, the pressure drop may be computed by using equations 4, 3, 2, and 1. In this method, the numerical constant of equation 5 need not be evaluated. This is left as an exercise to the reader.

2-3.

Since the ratio of upstream pressure to downstream pressure is high(>2), sonic flow is anticipated. One method of solving this problem is to balance energy along the flow path for an adiabatic flow as follows.

$$\frac{(k+1)}{2}\ln\frac{2+(k-1)N_{Ma1}^2}{(k-1)N_{Ma1}^2} - \left(\frac{1}{N_{Ma1}^2}-1\right)+\gamma(4fL/D)=0$$

Where:

k = Ratio of specific heat of gas at constant pressure to the specific heat of the same gas at constant volume.

N_{Ma} = ratio of the gas velocity to the velocity of sound in the gas under the same pressure and temperature(Mach number).

f = Fanning friction factor.

L = Length of pipe of internal diameter D. L and D must have the same unit.

Subscript 1 indicates upstream condition.

Step 1: Calculate the Fanning friction factor.

$f = 0.0625[\log(3.7x1.049/(0.00015x12))]^{-2} = 0.00562$

Step 2: Substitute the values of the known parameters and solve for the Mach number at the upstream condition by trial and error using the above equation.

$$\frac{(1.4+1)}{2}\ln\left(\frac{2+(1.4-1)N_{Ma1}^2}{(1.4+1)N_{Ma1}^2}\right) - \left(N_{Ma1}^{-2}-1\right)+1.4\times4\times0.00562\times25\times12/1.049 = 0$$

$$1.2\ln\left(\frac{2+0.4N_{Ma1}^2}{2.4N_{Ma1}^2}\right) - N_{Ma_1}^{-2}+10=0$$

By trial and error, $N_{Ma1} = 0.278$

Step 3: Calculate factor Y_1:

$Y_1 = 1 + (\gamma-1)(N_{Ma1}^2)/2 = 1 + (1.4 - 1)(0.278^2)/2 = 1.0155$

Step 4: Calculate T_{choked}:

$T_{choked} = 2Y_1T_1/(\gamma+1)$

where:

T_{choked} = Critical flow temperature in absolute scale
T_1 = Upstream temperature in absolute scale

$T_{choked} = 2x1.0155x(460+90)/(1.4+1) = 465.4\ ^oR$

Step 5: Calculate P_{choked}:

$P_{choked} = P_1 N_{Ma1}[2Y_1/(\gamma+1)]^{0.5}$
where:
P_{choked} = Critical flow pressure in absolute scale
P_1 = Upstream pressure in absolute scale

$P_{choked} = 114.7x0.278x[2x1.0155/2.4]^{0.5} = 29.33$ psia = 4223.96 lbf/ft^2

Since the pressure at the top of distillation column is 5 psig or 19.7 psia(<29.33), the critical flow is confirmed.

Step 6: Calculate G_{choked}:

$G_{choked} = P_{choked}[\gamma g_c M/(RT_{choked})]^{0.5}$
where:
G_{choked} = Critical mass velocity, $lb/ft^2.s$
$g_c = 32.2\ ft.lb/lbf.s^2$
M = Molecular weight, lb/lbmol
R = 1545 ft.lbf/lbmol.^{o}R

$G_{choked} = 4223.96x[1.4x32.2x28/(1545x465.4)]^{0.5}$
$= 176.98\ lb/ft^2.s$

Step 7: Calculate flow rate:

Flow rate = Mass velocity x internal cross section of pipe = $176.98x\pi x(1.049/12)^2/4 = 1.062$ lb/s

2-4.1
The linear velocity through an orifice may be computed by(equation 2-17 of CELR):

$$u_o = C_o Y\left(\frac{2g\Delta H}{1-\beta^4}\right)^{0.5} ft/s$$

Since liquid is incompressible, Y = 1. Since the ratio of the diameter of the orifice to the diameter of the tank is very small, β^4 may be assumed as 0. So, the above equation reduces to:

$u_0 = C_0(2g\Delta H)^{0.5}$ ft/s

If A_0 is the cross section of the orifice in ft^2, A is the cross section of the tank in ft^2, and H is the height of the liquid level above the center of the orifice in ft at any instant t, then the instantaneous volumetric flow rate may be given by:

$$-A(dH/dt) = A_0 u_0 = A_0 C_0 (2g\Delta H)^{1/2}$$

(Please note that the negative sign implies a fall in level with time)

Integrating the above equation one gets:

$$t = \frac{A(2/g)^{1/2}}{A_0 C_0}\left(H_i^{1/2} - H_f^{1/2}\right), \text{ seconds}$$

From the given data:

$A = \pi(10)^2/4 = 78.54\ ft^2$
$g = 32.2\ ft/s^2$
$A_0 = \pi(1.049/12)^2/4 = 0.006\ ft^2$
$C_0 = 0.61$ assumed
$H_i = 20 - 1 = 19$ ft
$H_f = 5 - 1 = 4$ ft

Substituting in the above equation:

$$t = \frac{78.54 \times (2/32.2)^{1/2}}{0.006 \times 0.61}\left(19^{1/2} - 4^{1/2}\right)$$
$$= 12615.6 \text{ s}$$
$$= 3.5 \text{ h}$$

Discussion:

If the above procedure is used to calculate the draining time to empty the tank ($H_f = 0$) or to calculate the draining time for high viscosity liquid, the calculated time will be approximate(lower than real time). This is because C_0 is a function of Reynold's number based on velocity through the orifice. The value of C_0 at low Reynold's number may be approximately 0.2 which assymptotically increases to 0.61 at very high Reynold's numbers. An accurate estimation of draining time requires a plot of C_0 against Reynold's number based on flow through the orifice, and graphical integration. This is left as an exercise to a zealous reader.

2-4.2

For the second part of the problem, the draining time,t, in seconds may be computed by:

$$t = \frac{(D_T)^2(2g_c)^{1/2}}{D^2 g}[1 + \Sigma K + 4f\Sigma(L/D)]^{1/2} \text{ X}$$

$$[\{\Delta P/\rho + H_i(g/g_c)\}^{1/2} - \{\Delta P/\rho + H_f(g/g_c)\}^{1/2}]$$

2-10

Data available:

D_T = Inside diameter of tank, ft = 10 ft

g_c = 32.17 lb.ft/lbf.s^2

D = Inside diameter of pipe, ft = 1.049/12

= 0.08742 ft

g = 32.17 ft/s^2

ΣK = Sum of K-factors for entrance and exit= 1.5

f = Fanning friction factor = $0.0625[\log_{10}(3.7D/\epsilon)]^{-2}$, assumed

where ϵ = roughness factor of pipe = 0.00015 ft, assumed for metallic pipe.

f = $0.0625[\log_{10}(3.7x0.08742/0.00015)]^{-2}$ = 0.005624

$\Sigma(L/D)$ = Sum of L/D-factors for straight pipe and other fittings whose resistances have not been included in the K-factors above = 10/0.08742 = 114.39

ΔP = Pressure on the liquid surface in vessel,lbf/ft^2 - pressure at the point of discharge,lbf/ft^2 = 0, because both are atmospheric.

H_i = Initial height of liquid from the center of the pipe = 20 - 1 = 19 ft

H_f = Final height of liquid from the center of the pipe = 5 - 1 = 4 ft

Substituting the above numerical values:

$$t = \frac{10^2(2x32.17)^{1/2}}{(0.08742)^2x32.17}[1 + 1.5 + 4x0.005624x114.39]^{1/2}[19^{1/2} - 4^{1/2}]$$

= 17334.975 s

= 4.8 h

HEAT TRANSFER

3-1.

Liquified petroleum gas is transported in a spherical uninsulated tank fitted with a relief valve required by law to prevent overpressurization of the vessel. The owner realizes that he is losing the gas through the valve due to the transmission of heat from the surrounding and consequent vaporization, and decides to install insulation on the tank. Given the following data, make an approximate estimate of the economic optimum thickness of insulation.

Tank outside diameter: 3.65 m
Average temperature inside the tank: 90 K
Latent heat of vaporization: 0.25×10^6 J/kg
Thermal conductivity of insulation: 0.035 W/m.K
Depreciation and maintenance cost of insulation: $15.00/yr.$m^3$
Value of the liquified gas: $0.10/kg
Average ambient temperature: 300 K
Tank remains full for 500 hours/year.

3-2.

The fouling study of a water cooled, liquid-to-liquid heat exchanger indicates that the overall fouling for the application follows the Kern-Seaton model as follows:

$$R_f = R_f^*[1 - \exp(-t/t_c)]$$

Where:

R_f = Overall fouling resistance, h.ft^2.°F/Btu, at a time t months after operation starts.
R_f^* = Asymptotic fouling resistance, h.ft^2.°F/Btu. This is the maximum fouling resistance reached after the critical time period t_c months.

The fouling resistance is predominantly on the water side. For this application R_f^* = 0.002 h.ft^2.°F/Btu and t_c = 16 months. The plant practice requires continuous operation of 11 months and shut down for 1 month for maintenance and cleaning. The inlet temperature, physical properties at the inlet conditions, and flow rate of the process fluid are constant. The outlet temperature of the process fluid is to be controlled by manipulating the cooling water flow rate. The inlet temperature of cooling water is constant. Assume that LMTD of the fouled exchanger at the end of the operation cycle(11 months) is 1.25 times the LMTD of the clean exchanger at the beginning of the operation. Find the required overall heat transfer coefficient of the exchanger under clean condition.

3-2

3-3.

The following data are available for a shell and tube heat exchanger.

	Shell Side	Tube Side
Liquid	Process	Water
Flow, lb/h	50,000	Variable
Temperature, in °F	175	90
Temperature,out °F	130	Variable
Sp. heat, Btu/lb.°F	0.8	1

The overall clean heat transfer coefficient is 250 Btu/h.ft^2.°F. The cooling water contains soluble calcium carbonate and calcium sulfate from pH control. Use your judgement to limit the outlet temperature of cooling water. The plant practice requires 3/4" ODx 0.62" ID tubes laid out on 15/16", 60° triangular equilateral pitch. The minimum tube side velocity is given by[1]:

$$R_f^* = 0.006481/v^{1.33}d_i^{0.23}$$

Where:

R_f^* = Asymptotic fouling resistance, = 0.002 h.ft^2.°F/Btu

v = minimum recommended tube side velocity, ft/s

d_i = Inside tube diameter, in

Because of space limitation, the overall length of the exchanger must be less than 15 ft.

3-3.1

The heat duty of the heat exchanger is:

(A) Variable (B) Gradually decreasing (C) Gradually increasing (D) 1.8x10^6 Btu/h

3-3.2

If the outlet temperature of cooling water is limited to 120°F, the LMTD of a true counter-current exchanger(°F) is:

(A) 47.5 (B) 35 (C) 298.3 (D) 47.1

3-3.3

For a clean exchanger with an LMTD of 47.1°F, the area(ft^2) for required duty is:

(A) 153 (B) 300 (C) 75 (D) 100

3-3.4

If the cooling water outlet temperature is allowed to rise to 120°F, the cooling water flow rate(lb/h) is:

(A) 30,000 (B) 15,000 (C) 60,000 (D) 12,000

3-3.5

The minimum recommended tubeside velocity(ft/s) is close to:

(A) 5.3 (B) 8.6 (C) 2.6 (D) 4.1

3-3.6

If the minimum recommended tubeside velocity is taken as 2.63 ft/s, the number of tubes per pass for a cooling water flow rate of 60000 lb/h under clean condition is approximately:
(A)200 (B)100 (C) 25 (D) 48

3-3.7

If 48 tubes per pass is assumed for design, and total area required is 152.9 ft^2, the the number of tube passes is:
(A)1 (B)4 (C)2 (D)6

3-3.8

Assuming 1-2 shell-and-tube design with a maximum cooling water outlet temperature of 120°F, the correction factor for LMTD is close to:
(A) 1 (B) 1.5 (C) 0.75 (D) .9

3-3.9

Assuming 1-2 shell-and-tube design with a maximum cooling water outlet temperature of 120°F, and a correction factor for LMTD as 0.89, the recommended tube length(ft) is:
(A) 4 (B) 6 (C) 2 (D) 10

3-3.10

Assuming 1-2 shell-and-tube design and 48 tubes per pass, the recommended nominal shell diameter in inch is:
(A)24 (B)48 (C)6 (D)12

Estimate the area, tube length, shell diameter, number of shell and tube passes for the above application.

3-4.

The following data are available for a shell and tube heat exchanger.

	Shell Side	Tube Side
Liquid	Process	Water
Flow, lb/h	50000	60000
Temperature,in °F	175	90
Temperature,out °F	130	120
Sp. heat,Btu/lb.°F	0.8	1
Viscosity, cP	1.5	0.9
Th. conductivity Btu.ft/h.ft^2.°F	0.37	0.36

The required overall heat transfer coefficient under clean condition is 250 Btu/h.ft^2.°F. The mechanical data are as follows.

Shell diameter: 12 in (internal)
Tubes: 3/4" ODx0.62" ID x 10' long

3-4

Tube pitch: 15/16",60° equilateral, triangular
Number of tubes: 96
Number of tube passes: 2
Number of shell passes: 1

The flow rate, inlet temperature of the process side(shell side) fluid and inlet temperature of water will be constant. The outlet temperature of process fluid will be held constant by manipulating the water flow rate. The water flow rate shown is under clean condition of the heat exchanger. Estimate the number of baffles, baffle pitch, LMTD, cooling water flow, and water velocity under fouled condition at the end of the operation cycle when the fouling resistance reaches 0.001 h.ft^2.°F. The thermal conductivity of tube material is 30 Btu.ft/h.ft^2.°F.

SOLUTION

3-1.

Equation (3-19) from CELR may be used with the mean heat transfer area of insulation from equation (3-20) for spherical vessel. For an approximate estimate, the resistances of inside vapor film, metal wall, and outside air film and radiation may be ignored. Thus:

$$q = \frac{t_s - t_a}{x_i/(k_b A_{mb})} \qquad \text{Equation (1)}$$

where:

q = heat transfer rate, W
t_s = Inside temperature of vessel, K
t_a = Ambient temperature, K
x_i = Insulation thickness, m
= $(D_s - D_o)/2$
D_s = Diameter of vessel across the insulation, m
D_o = Outside diameter of uninsulated vessel,m
k_b = Thermal conductivity of insulation,W/m.K
A_{mb} = Mean area of insulation, m^2

$=(\pi D_s^2 \pi D_o^2)^{1/2}$
$= \pi D_s D_o$

Substituting x_i and A_{mb} in above equation:

$$q = \frac{2\pi k_b D_s D_o (t_s - t_a)}{D_s - D_o} \text{, J/s} \qquad \text{Equation (2)}$$

Loss of gas due to vaporization = q/L, kg/s Equation (3)
where L = Latent heat of vaporization, J/kg

If c_g = Annual cost of lost gas,
T = hours of operation per year
c_1 = cost of gas, \$/kg
$c_g = 3600qc_1T/L$

$$q = \frac{(2)(3600)\pi k_b D_o c_1 (t_s - t_a) T D_s}{L\,(D_s - D_o)}$$

$= KD_s/(D_s - D_o)$, \$/yr Equation (4)

where $K = 7200\pi k_b D_o c_1 (t_s - t_a)T/L$ Equation (4a)

K is a constant by the conditions of the problem.

Volume of insulation = $\pi(D_s^3 - D_o^3)/6$, m^3
If d = annual depreciation and maintenance cost of insulation, \$/m^3.yr, and c_i is the annual cost of insulation, \$/yr, then:

$c_i = \pi d(D_s^3 - D_o^3)/6$, \$/yr Equation (5)

If c_T = total annual cost, then:
$c_T = c_g + c_i$
$= KD_s/(D_s - D_o) + \pi d(D_s^3 - D_o^3)/6$ Equation (6)

Differentiating c_T with respect to D_s and equating the derivative to zero, one gets:

$D_s = [D_o + (D_o^2 + 4m)^{1/2}]/2$ Equation (7)

Where $m = (2D_oK/\pi d)^{1/2}$ Equation (8)

(Note: To prove that D_s thus obtained is for the minimum cost, one has to prove that the second derivative of c_T with respect to D_s is positive. This is left as an exercise to the reader. Equation (7) ignores the negative determinant because it lacks physical significance.)

The optimum thickness of insulation, t_{opt} may be given by:

$t_{opt} = (D_s - D_o)/2 = [(D_o^2 + 4m)^{1/2} - D_o]/4$ Equation (9)

Substitute the numerical values of the parameters in equation (4a).

$K = 7200x\pi x0.035x3.65x0.10x210x500/0.25x10^6 =$ 121.36

3-6

Substuting the value of K thus obtained, and the numerical values of $D_o = 3.65$, $d = 15$ in equation (8) one gets:

$$m = (2x3.65x121.36/\pi x15)^{1/2} = 4.336$$

Substituting the values of m and D_o in equation (9), one gets:

$$t_{opt} = [(3.65^2 + 4x4.336)^{1/2} - 3.65]/4 = 0.472 \text{ m}$$

3-2.

The basic equation for heat balance in a heat exchanger is:

$Q = UAF(\Delta T)_m$ Equation (1)

Where:

Q = heat exchanged, Btu/h
U = Overall heat transfer coefficient, Btu/h.ft^2.°F
A = area of heat exchange, ft^2
F = correction factor for deviation from true counterflow and temperature cross.
$(\Delta T)_m$ = Log-mean temperature difference between the fluid, °F

By the statement of the problem, Q is to be maintained constant. For an installed exchanger, A is constant. The parameters that will vary over time are U, F, $(\Delta T)_m$. In the beginning, the controller will throttle water control valve thereby increasing the water outlet temperature and consequently decreasing the LMTD. As the exchanger gets fouled, water flow rate will be increased, and so will LMTD to compensate for the fouling. Thus:

$Q = U_cAF_c(\Delta T)_{mc} = U_fAF_f(\Delta T)_{mf}$ Equation (2)

Where the subscript c denotes clean condition, and f denotes fouled condition. Since R_f is the overall fouling resistance, we can write:

$1/U_f = 1/U_c + R_f = (1 + R_fU_c)/U_c$
or, $U_f = U_c/(1 + R_fU_c)$ Equation (3)

Assuming $F_c = F_f$, one gets from equations 2 and 3:

$U_c(\Delta T)_{mc} = U_f(\Delta T)_{mf} = U_c(\Delta T)_{mf}/(1 + R_fU_c)$
$U_c = [(\Delta T)_{mf}/(\Delta T)_{mc} - 1]/R_f$ Equation (4)

By the statement of the problem:

$(\Delta T)_{mf}/(\Delta T)_{mc} = 1.25$

Substituting the values of $R_f^* = .002$, $t=11$, and $t_c=16$ in the given equation:

$$R_f = 0.002[1 - \exp(-11/16)] = 0.001$$

Substituting these values in equation 4:

$U_c = (1.25 - 1)/0.001 = 250$ Btu/h.ft^2.°F

3-3.

Based on the salt content of water and the outlet temperature of process fluid, the outlet temperature of cooling water should not exceed 120 °F for a preliminary design. As the exchanger is fouled, more water will be called in by the controller, and the outlet temperature of water will drop.

3-3.1

From the data of shell side fluid:

$Q = 50{,}000$ (lb/h) x 0.8 (Btu/lb.°F) x (175 - 130)(°F)
$= 1.8\text{x}10^6$ Btu/h **Answer is D.**

3-3.2

By limiting the outlet temperature of the cooling water to 120 °F, the LMTD may be calculated for the clean heat exchanger.

```
175 ------------------> 130
120 <-----------------   90
 55                      40
```

$(\Delta T)_{mc} = (55 - 40)/\ln(55/40) = 47.1$ °F **Answer is D.**

3-3.3

The area needed may be computed by:

$$A = \frac{Q}{U_c F_c (\Delta T)_{mc}}$$

Where:

$Q = 1.8\text{x}10^6$ Btu/h as computed above.
$U_c = 250$ Btu/h.ft^2.°F, given.
F_c is unknown since the geometry is unknown, assume 1.
$(\Delta T)_{mc} = 47.1$ °F as computed above.

Substituting these values,

$A = (1.8\text{x}10^6)/(250\text{x}47.1) - 152.9$ ft^2 **Answer is A.**

3-3.4

By heat balance with cooling water, the water flow rate(m_c) may be obtained from:

$m_c = Q/(C_{pc}.\Delta T) = 1.8\text{x}10^6/(120\text{-}90) = 60{,}000$ lb/h **Answer is C.**

3-3.5

Before estimating the number of tubes and tube-passes, the tube side velocity has to be determined. This may be done from:

$R_f^* = 0.006481/v^{1.33} d_i^{0.23}$

where:
R_f^* = Asymptotic fouling resistance = 0.002, given.
d_i = 0.62 in given in Problem 3.

Substituting these values,
v = 2.63 ft/s (minimum velocity) **Answer is C.**

3-3.6
If N_t = number of tubes per pass, then:

$N_t x(\pi/4)x(0.62/12)^2(ft^2)x2.63(ft/s)x3600(s/h) x61.9(lb/ft^3) = 60{,}000$ lb/h
Solving for N_t:
N_t = 48.8 tubes per pass.
To avoid a velocity lower than the minimum velocity, assume N_t = 48. **Answer is D.**

3-3.7
If L = length of tubes, ft
n = number of tube passes

Then:
$L(ft)x(\pi x0.75/12)(ft^2/tube.ft)x48(tubes/pass)\ xn(pass) = 152.9\ ft^2$

nL = 16.2

Because of the space limitation, assume n = 2 to give tube length, L = 8.1 ft. **Answer is C.**

3-3.8
Assuming 1 shell pass, we can now estimate the LMTD correction factor under clean condition.

Referring to Kern[2a]

$R=(T_1 - T_2)/(t_2 - t_1)=(175 - 130)/(120 - 90)=1.5$
$S=(t_2 - t_1)/(T_1 - t_1)=(120 - 90)/(175 - 90)=0.35$

Corresponding to these values, F_c = 0.89. **Answer is D.**

3-3.9
Allowing for this correction factor, the tube length should be 8.1/.89 = 9.1 ft. Suggested tube length is 10 ft. **Answer is D.**

3-3.10
Referring to Table 9 of Kern[2b], a 12 in ID shell would fit 98 tubes. Suggested design:

Shell diameter : 12 in (internal)
Tubes: 3/4" ODx 0.62" ID x 10 ft long

Tube pitch: 15/16",60° equilateral, triangular
Number of tubes = 96
Number of tube passes = 2
Number of shell passes = 1
Surface area = 188.5 ft^2 **Answer is D.**

3-4.

The solution of this problem involves the estimation of shell side film coefficient from the estimated tube side coefficient and the overall heat transfer coefficient.
The tube side coefficient may be estimated by equation (3-33a) of (CELR):

$$h_i = 0.023(D_iG/\mu_b)^{0.8}(c_p\mu/k)^{0.4}{}_b(k/D_i)$$

Where:

D_i = tube diameter = 0.62/12 = 0.05167 ft
Total tube side flow area
= number of tubes per pass x cross section of one tube
= $(96/2)x(\pi/4)x(0.05167)^2 = 0.10064$ ft^2
G = tube side mass velocity = water flow(lb/h)/ tube side flow area(ft^2)
= 60,000/0.10064 = $5.96184x10^5$ lb/h.ft^2
μ_b = tube side viscosity = 0.9 cP = 0.9x2.42 = 2.178 lb/ft.h
c_p = tube side specific heat = 1 Btu/lb.°F
k = tube side thermal conductivity = 0.36 Btu.ft/h.ft^2.°F

Substituting these numerical values:

h_i=0.023x$(0.05167x5.96184x10^5/2.178)^{0.8}$ x$(1x2.178/0.36)^{0.4}$x(0.36/0.05167)
= 688.6 Btu/h.ft^2.°F

From equation (3-27) of (CELR), ignoring the fouling resistance under clean condition:

$$1/U_o = (1/h_i)(D_o/D_i) + (l_w/k_w)(D_o/D_{av}) + 1/h_o$$

U_o = 250 Btu/h.ft^2.°F, given.
D_{av} = (0.75 - 0.62)/ln(0.75/0.62) = 0.68293 in
1/250 = (1/688.6)x(0.75/0.62)+ ((0.75-0.62)/24x30)x(0.75/0.68293)+ $1/h_o$

Hence, h_o = 489 Btu/h.ft^{2o}.F

The baffle spacing should be such that above shell side coefficient may be achieved.
Ignoring the viscosity correction factor, the shell side coefficient may be given by equation (3-59) of (CELR):

$$(h_oD_e/k) = 0.36(D_eG_s/\mu)^{0.88}(c_p\mu/k)^{1/3} \quad \text{Equation(1)}$$

From equation (3-62) of (CELR):

$D_e = (0.86P_T^2 - (\pi/4)d_o^2)/(3\pi d_o)$, ft

Where:

P_T = tube pitch = 15/16" = 0.9375 in
d_o = tube OD = 0.75"
$D_e = (0.86x(0.9375)^2 -(\pi/4)x(0.75)^2)/(3x\pi x0.75)$ = 0.04443 ft

Substituting in equation (1):

$(489x0.04443/0.37)=0.36(0.04443G_s/(1.5x2.42))^{0.55}x(0.8x1.5x2.42/0.37)^{1/3}$

Solving for G_s:

G_s = 246,968 lb/h.ft^2

Also:

3-10

$G_s = W/a_s$

Where:

W = Shell side fluid flow = 50000 lb/h

$a_s = (D_{si}C'B)/(144P_T)$, ft^2

Where:

D_{si} = Shell ID = 12 in

C' = clearance between tubes = 15/16 - 3/4
= 0.1875 in

B = Baffle spacing(in), unknown

P_T = 15/16 = 0.9375 in

a_s = (12x0.1875B)/(144x0.9375)
= 0.016667B

Therefore,

G_s = 50000/0.016667B = 246968

Hence,

B = 12.147 in

Assume B = 12 in, since maximum recommended value of B is D_{si}

Number of baffles = Tube length(in)/B - 1 = 10x12/12 - 1 = 9

The overall heat transfer coefficient under fouled condition at the end of the operation cycle, U_{of}, may be given by:

$1/U_{of} = 1/U_o + R_f$ = 1/250 + 0.001

Hence, U_{of} = 200 Btu/h.ft^2.°F

The LMTD under fouled condition may be obtained from:

$(\Delta T)_{mf} = Q/(AU_{of}F_f)$ = (1.8x10^6)/(200x188.5x0.89) = 53.7 °F

Now estimate the outlet temperature of cooling water.

175 --------------------->	130
t_2 <---------------------	90
175-t_2	40

LMTD = (135-t_2)/ln((175-t_2)/40) = 53.7

By trial and error, t_2 = 104.8 °F.

Estimate the cooling water flow and velocity at the end of operation cycle.

Cooling water flow rate = 1.8x10^6/(104.8 - 90) = 121621.6 lb/h

Water velocity through tubes =121621.6/(48x(π/4)x(0.62/12)^{2}x3600x61.9) = 5.42 ft/s

References:

1. AIChE Today Series: Fouling of Heat Transfer Equipment. American Institute of Chemical Engineers, 345 E 47 Street, New York, NY 10017.

2. D. Q. Kern Process Heat Transfer, McGraw-Hill, New York, 1950:(a) p828 (b) p842

EVAPORATION

4-1: A solution is to be concentrated from 10 to 65 % solids in a vertical long tube evaporator. The solution has a negligible boiling point elevation, and its specific heat is 0.97 kcal/kg-K. Steam is available at 1.07 kg/cm^2 G. The evaporator is to operate at 0.138 kg/cm^2 A. The feed enters the evaporator at 303 K. The total evaporation is to be 22000 kg/h of water. Calculate the heat to be transferred and the steam economy. The overall heat transfer coefficient is 2400 kcal/h-m^2-C.

4-2: A double-effect forward feed evaporator is required to concentrate a solution of NaOH from 10 to 50 % by wt. The heat transfer coefficients are : first effect U = 450 Btu/h-ft^2-^{0}F. , second effect U = 250 Btu/h-ft^2-^{0}F. 60000 lb/h of solution enters the evaporator at 120 ^{0}F. and steam at 15 psig is available. The vacuum in the second effect is held at 28 in Hg. Water from a cooling tower is supplied at 86 ^{0}F. Estimate (a) Live steam and condenser water amounts required (b) heat transfer surface per unit. and (3) steam economy.

4-3: It is proposed to install a multi-effect evaporation system to concentrate a solution which involves evaporation of 70000 lb/h of water. A senior process engineer studied the problem and tabulated his results as given in Table 4-1. Recommend your selection based on the following cost data: State the assumptions you make. Assume on stream factor of 8000 h/yr.

Steam (low pressure) - \$2.86/Mlb
Cooling water - \$0.1/Mgal
Electricity - \$0.07/ KWh
One operator per shift - \$10 per hour
Annual cost of tube renewal :

No. of effects	4	5	6	7
cost /yr \$	2100	2000	2300	2630

Make any additional assumptions as necessary.

Table 4-1: Summary of evaporator calculations

Number of effects	4	5	6	7
Investment \$	305000	374500	423100	476000
Hourly evaporation, lb/h	70,000	70,000	70,000	70,000
Steam usage, lb/h	22,000	18,000	15,300	13,400
Condenser water, gpm	1,388	1,145	1,014	895
Electricity, Hp	20	25	30	40
Man-hours per day	24	24	24	24

SOLUTIONS

4-1:

Temperature of water at 0.138 kg/cm^2 A = 325 K
Steam temperature at 1.07 kg/cm^2 G = 394 K
Latent heat of vaporization at 325 K = 567.6 kcal/kg
Material balance: let F = feed and L = thick liquor

Solute balance : F(0.1) = L(0.65) Therefore F = 6.5 L
By water balance F (0.9) = V + L (0.35) where V = required evaporation = 27000 kg/h.
substituting appropriate values in the equation gives 0.9 x 6.5 L = 27000 + L(0.35)
From which L = 4909 kg/h and then F = 31909 kg/h
Use boiling point in the evaporator as the datum temperature.
Then Q = 27000(567.6) + 31909(0.97) (325 - 303) = 16.006x10⁶ kcal/h

T = 394 - 325 = 69 K

Heat transfer area = $\frac{16.006\times10^6}{2400\times69}$ = 96.7 m^2

λ_S = 525.3 kcal/kg

Therefore steam required = (16.006×10^6)/525.3 = 30470 kg/h
and steam economy = 27000/30470 = 0.89 kg/kg of steam.

4- 2:

Vacuum in second effect = 28 in Hg = 29.92 - 28 = 1.92 in Hg pressure = 0.94 psia.
From steam tables, at 0.94 psia, t = 98 °F , H_S = 1104.2 Btu/lb, h_C = 66 Btu

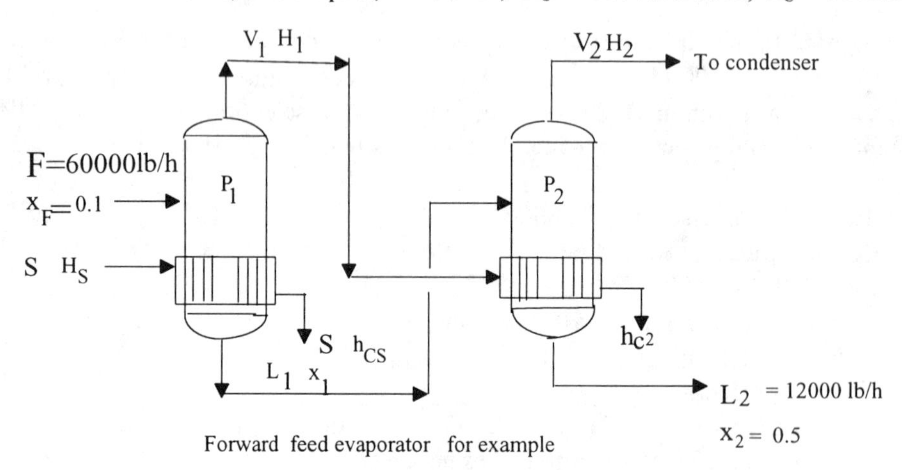

Forward feed evaporator for example

Material balance:

Solute = 0.1 x 60000 = 6000 lb/h NaOH
Assume evaporation in the first effect = 26000 lb/h.
Therefore, evaporation in 2nd effect = 22000 lb/h
L_1 = 60000 - 26000 = 34000 lb/h x_1 = 6000/34000 = 0.1765 h_F = 80 btu/lb
L_2 = 34000 - 22000 = 12000 lb/h x_2 = 0.5
Assume BP of water in first effect = 215 °F.
BP of NaOH solution = 226 °F (from figure)
Boiling point of water in 2nd effect = 98 .
BP of NaOH solution = 171 °F (from figure)
BPR in first effect = 226 - 215 = 11°F
BPR in 2nd effect = 171- 98 = 73 °F
Overall temperature difference available = 250 - 98 - 11-73 = 68 °F.
First estimate of temperature differences :

U_1 = 450 Btu/h-ft²-°F. U_2 = 250 Btu/h-ft²-°F

$$\frac{\Delta T_1}{\Delta T_2} = \frac{U_2}{U_1} = (250/450) = 0.5556$$

$$\Delta T_1 + \Delta T_2 = 68\ ^0F$$

Then $\frac{\Delta T_1}{\Delta T_1 + \Delta T_2} = \frac{0.5556}{1+0.5556} = 0.357$

$\Delta T_1 = 0.3572 \times 68 = 24$ ^oF $\Delta T_2 = 68 - 24 = 44$ ^oF

Prepare temperature frame as follows

	Temp. ^oF	λ Btu/lb	Superheat ^oF	Superheat + λ Btu/lb	h - liq. enthalpy Btu/lb	H- Vapor enthalpy Btu/lb
Steam to I	250	945.2	0	945.2		1164.1
ΔT_1	24					
Boiling point in I	226				171	
Boiling point rise	11					

	Temp.	λ	Superheat	Superheat + λ	h - liq. enthalpy	H- Vapor enthalpy
Saturation temperature of vapor to II	215	968.4	5	973.4		1156.6
ΔT_2	44					
Boiling point in II	171				202	
BPR in II	73					
Saturation temperature to condenser	98	1038	33	1071		1137.2

First effect:

80(60000) + 945.2 (S) = 26000(1156.6) + 171(34000)
Which gives S = 32888 lb/h steam to first effect

$A_1 = \frac{945.2\times32888}{450\times24} = 2878\ ft^2$ $A_2 = \frac{973.4\times26000}{250\times44} = 230\ ft^2$

These areas are unequal. Therefore, a second trial is required. For this trial redistribute the temperature differences based on areas as follows:
Average calculated area = (2878 + 2301)/2 = 2589.5 ft^2
Then ΔT_1 = (2858/2589.5) x 24 = 26.0. ^oF ΔT_2 = (2301/2589.5) x 44 = 39 ^oF
Then boiling point in 1st effect = 250 - 26 = 224 ^oF
Total boiling point rise = 250 -98 - 65 = 87 ^oF
BPR in 2nd effect = 73 ^oF therefore BPR in first effect = 87 - 73 = 14 ^oF
Now prepare another temperature frame as follows

	Temp. ^oF	λ Btu/lb	Superheat ^oF	Superheat + λ Btu/lb	h - liq. enthalpy Btu/lb	H- Vapor enthalpy Btu/lb
Steam to I	250	945.2	0	945.2		1164.1
DT_1	26					
Boiling point in I	224				168	
Boiling point rise	14					
Saturation temperature of vapor to II	210	971.9	6 .3	978.2		1155.9
DT_2	39					
Boiling point in II	171				202	
BPR in II	73					
Saturation temperature to condenser	98	1038	33	1071		1137.3

$S(945.2) + 80(60000) = V_1(1155.9) + 168(60000-V_1)$

$V_1(978.2) + (60000 - V_1)168 = (48000 - V_1)(1137.3) + 12000(202)$

From last equation, $V_1 = 24100$ lb/h

Then from first equation $S = 30775$ lb /h

$$A_1 = \frac{945.2 \times 30775}{26 \times 450} = 248 \text{ ft}^2 \qquad A_2 = \frac{24100 \times 978.2}{39 \times 250} = 2418 \text{ ft}^2$$

These value are close enough. $A = (2486 + 2418)/2 = 2452$ say 2500 ft²

Steam economy = 48000/30775 = 1.56 lb/lb of steam.

4-3:

Assumptions made are as follows : Annual depreciation 10 % straight line method. 10 yr life.
Interest at 5% Maintenance at 1.5% Labor at \$10/h
on stream hours = 8000 h/yr

No. of effects	4	5	6	7
Investment	305000	374500	423100	476000
Incremental investment	0	69500	118100	171000
Cost of steam @ 2.86/Mlb	503360	411840	350064	306592
Cost of water @ \$0.1/1000 gal	66624	54960	48672	42960
Cost of power @ 0.07/kwh	8579	10724	12869	21448
Labor @ \$10/h	87600	87600	87600	87600
Tube renewal \$	2100	2000	2300	2630
Maintenance @ 1.5 %	4575	5618	6347	7140
Annual depreciation @ 10 %	30500	37450	42310	47600
Interest @ 5%	15250	18725	21155	23800
Total annual expenditure \$	718588	628917	571322	539770
Incremental annual savings \$	0	89671	147266	178818
Ratio of annual incremental savings to incremental investment		1.29	1.246	1.045

The ratio of annual savings to incremental investment is highest for the quintuplet evaporator system and therefore , it should be installed.

DISTILLATION

5-1

5-1.1 Which one(s) of the following stages is(are) **not** considered an ideal stage in distillation concept?

Stage #	Feed Type	Product Type	Other Feature
1	Vapor & liquid	Vapor thru top Liquid thru bottom	
2	Vapor	All liquid	Stage condenser
3	Vapor	Vapor thru top Liquid thru bottom	Partial Condenser
4	Liquid	Vapor thru top Liquid thru bottom	Reboiler

Wherever there are two phases, assume that the phases are perfectly mixed and are in equilibrium with each other, and also assume steady state conditions.
(A)1 (B)2 (C)3 (D)4

5-1.2 A liquid mixture of three components A,B,C having different relative volatilities are to be separated by ordinary distillation only into three pure components(sharp separation). What is the least number of theoretically possible sequences for this separation?
(A)2 (B)1 (C)3 (D)4

5-1.3 The overhead of a distillation column has the following relationship of pressure with temperature:

Temperature ,°F	Bubble-point pressure, psia	Dew-point pressure, psia
85	255	185
95	265	205
100	280	225
115	315	265
125	355	300

The cooling water available for condensing the overhead product has the summer time maximum temperature at 100 °F as it comes out of the cooling tower. If the approach of the reflux temperature to cooling water inlet temperature to condenser is 15 °F, the top pressure(psia) of the column should be:

(1)for a partial condenser: (A)280 (B)225 (C)315(D)265
(2)for a total condenser: (A)265(B)315(C)225(D)280

5-1.4 The following table shows the specification of the composition of the feed, distillate, and bottoms of a distillation column(the components are listed in the order of decreasing K-values):

Component	Feed(mol %)	Distillate(mol%)	Bottoms(mol%)
Methane	25	41.74	
Ethane	8	13.36	
Propane	26	43.11	0.44
n-Butane	18	1.79	42.21
n-Pentane	12		29.92
n-Hexane	11		27.43
	100	100	100

Based on the above specification,
(a) the light key is:
(A)Methane (B)Ethane (C)Propane (D)n-Butane
(b) the heavy key is:
(A)Propane (B)n-Butane (C)n-Pentane (D)n-Hexane

5-1.5 In problem 5-1.4, moles of distillate per 100 moles of feed are:
(A)100 (B)59.9 (C)67.9 (D) 20.7

5-1.6 In problem 5-1.4, total moles of n-butane in the bottoms per 100 moles of feed are:
(A)16.9 (B)21.1 (C)8.5 (D)10.5

5-1.7 In problem 5-1.4, $(\alpha_{LK/HK})_{av}$ = 1.98. Calculate the minimum number of theoretical stages including the reboiler as a stage.
(A)11.34 (B)3.95 (C)5.87 (D)17.95

5-1.8 In problem 5-1.4, if the specified separation requires 14 theoretical stages, then counting from the top, **estimate** the location of the feed tray.
(A)6 (B)3 (C)8 (D)9

5-1.9 The feed to a distillation column has 40% by weight of vapor, balance being liquid. What is the slope of the q-line with respect to the positive direction of the horizontal axis?
(A)1.5 (B)0.67 (C)-0.67 (D) -1.5

5-1.10 In problem 5-1.4, per 100 lbmoles of feed, the following data are available:
Feed enthalpy = -4.772×10^6 Btu/h, distillate enthalpy = -2.348×10^6 Btu/h, bottoms enthalpy = -2.476×10^6 Btu/h, reflux ratio = 5, and latent heat of overhead vapor = 5556.5 Btu/lbmol. Estimate the reboiler duty in million Btu/h per 100 lbmoles of feed.
(A)1.997 (B)1.945 (C)2.013 (D)2.105

5-2

Ammonia is to be distilled out of a still heated by steam through the jacket. The initial feed contains 2287 lbmoles of water and ammonia with ammonia concentration of 0.00314 mole fraction. If the ammonia concentration in the residue of the still is to be 0.00005 mole fraction, (a)how many moles of feed is to be distilled out? At this concentration, the ammonia-water equilibrium relationship may be given by $y = 13x$. (b)Assuming the feed at bubble point, and an average distillation rate of 200 lbmoles/h, estimate the distillation time.(c)Estimate the average concentration of ammonia in the distillate after the distillation is complete.(d) The vertical distance between the operating line and the equilibrium line of a theoretical stage of a distillation column is 10 mm. If the Murphree efficiency of the stage is 70%, what will be the corresponding distance for an actual stage?

5-3

The equlibrium relationship between two miscible, volatile components A,B(A=light key) is shown in Figure 5-1. A feed containing 50%(mol) of each component at its bubble point of 155 °F is to be separated by fractionation into an overhead stream with 95%(mol) light key and bottoms with 90%(mol) heavy key.(a) Using a reflux ratio of 1.25 of the minimum, calculate the number of theoretical stages required. (b)Assuming an average viscosity of the liquid as 0.5 cP, estimate the actual number of stages.(c)Assuming the distillate specific heat of 50 Btu/lbmol.°F, distillate

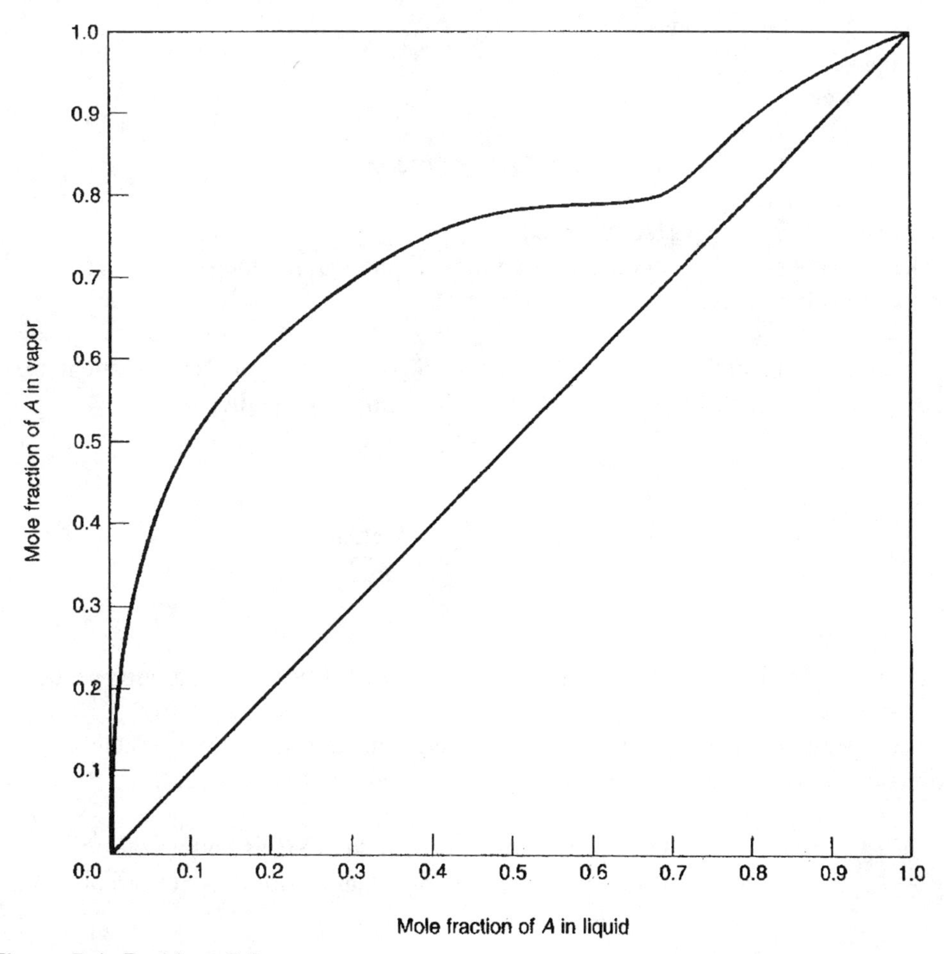

Figure 5-1. Problem 5-3.

temperature of 150 °F, overhead latent heat of 11000 Btu/lbmol, bottoms temperature of 160 °F, and bottoms specific heat of 51 Btu/lbmol.°F, calculate the reboiler duty per lbmol of feed.

5-4
A distillation column separating a binary mixture of components A & B has two feeds at bubble point. The following data are available:

Component	Relative Volatility	Feed Composition		Distillate Composition	Bottoms Composition
	α_{ij}	x_{F1}	x_{F2}	x_D	x_B
A(light key)	2.1	0.65	0.25	0.972	0.022
B(heavy key)	1.0	0.35	0.75		

Feed flows: F_1 = 105 moles/h, F_2 = 55 moles/h. Calculate the minimum reflux.

SOLUTIONS

5-1

5-1.1 **Stage #2**
Answer: B

Note:
An ideal distillation stage must have the following criteria:
(1)Must have vapor and liquid products.
(2)Vapor and liquid must be perfectly mixed.
(3)Vapor leaving the stage is in equilibrium with the liquid leaving the stage.
(4)Operation must be steady.

5-1.2 The number of theoretically possible sequences S_N of separating an N-component mixture into N pure-component products by **one** separation method only is given by:

$$S_N = \frac{[2(N-1)]!}{N!(N-1)!} = \frac{[2(2)]!}{3!2!} = 2$$

Thus, assuming that A,B,C are in the order of decreasing volatilities, the separation sequences could be:
(I) Sequence 1: Fractionate ABC into A as pure overhead, B&C as bottom products.
Fractionate B&C into B as pure overhead , and C as pure bottom product.

Or

(II) Sequence Fractionate ABC into A&B as overhead, and pure C as bottom product.
Fractionate A&B into pure A as overhead, and pure B as bottom product.

Answer: A

5-1.3

For partial condenser, column pressure = dew-point pressure corresponding to (t + a)
For total condenser, column pressure = bubble point pressure corresponding to (t +a)
where: t = maximum summer time cooling water temperature, a = approach.
Here t+a = 115 °F.
(1) Column pressure for partial condenser= 265 psia. **Answer: D**
(2) Column pressure for total condenser = 315 psia. **Answer: B**

5-1.4

By definition, the key components are the two components whose separation is specified in the distillation.
(a) The light key is propane. **Answer: C**
(b) The heavy key is n-Butane. **Answer: B**

5-1.5

Since both methane and ethane end up entirely in distillate, moles of distillate per 100 moles of feed = 25/0.4174 or 8/0.1336
=59.9 moles. **Answer: B**

5-1.6

Moles of bottoms = 12/0.2992 = 40.1
Moles of n-butane in distillate = 59.9x0.0179 = 1.1
Moles of n-butane in bottoms = 18 - 1.1 = 16.9
% n-butane in bottoms = 16.9/40.1= 42.1(check). **Answer: A**

5-1.7

$$N_{\min} = \frac{\log\left[\left(\frac{x_{LK}}{x_{HK}}\right)_D \left(\frac{x_{HK}}{x_{LK}}\right)_B\right]}{\log(\alpha_{LK/HK})_{av}}$$

From the data supplied:
Distillate: $x_{LK} = 0.4311$, $x_{HK} = 0.0179$
Bottoms: $x_{LK} = 0.44$, $x_{HK} = 0.4221$
$(\alpha_{LK/HK})_{av} = 1.98$
Substituting these values:

$$N_{\min} = \frac{\log\left[\frac{0.4311}{0.0179}\frac{0.4221}{0.0044}\right]}{\log(1.98)} = 11.34$$

Answer:(A)

5-6

5-1.8

The Underwood-Fenske equation:

$$\frac{N_R}{N_S} = \frac{\log\left[\left(\frac{x_D}{x_F}\right)_{LK}\left(\frac{x_F}{x_D}\right)_{HK}\right]}{\log\left[\left(\frac{x_F}{x_B}\right)_{LK}\left(\frac{x_B}{x_F}\right)_{HK}\right]}$$

Where:
N = number of theoretical stages
x = mole fraction of a key component
Subscripts: B = bottoms, D = distillate, F = feed, HK = heavy key, LK = light key, R = rectification section, S = stripping section.
From the given data,

$$\frac{N_R}{N_S} = \frac{\log\left[\frac{0.4311}{0.26}\,\frac{0.18}{0.0179}\right]}{\log\left[\frac{0.26}{0.0044}\,\frac{0.4221}{0.18}\right]} = 0.5706$$

$$\frac{N_R + N_S}{N_S} = \frac{N_T}{N_S} = 1.5706$$

$$N_S = \frac{N_T}{1.5706} = \frac{14}{1.5706} = 8.91$$

$$N_R = 14 - 8.91 = 5.09$$

Answer: A

5-1.9

$$\text{The slope} = \frac{q}{q - 1}$$

Where: q = fraction of liquid = 0.6.

$$\therefore \text{the slope} = \frac{0.6}{0.6 - 1} = -1.5$$

Answer:D

5-1.10

Overall energy balance around a distillation column:

$$Fh_F + Q_B = Dh_D + Bh_B + Q_C$$
$$Q_C = D(R + 1)\lambda$$

D= distillate per 100 lbmoles of feed from solution of problem 5-1.5 = 59.9 lbmoles, R=5, λ = 5556.5 Btu/lbmol.

$$Q_C = 59.9x6x5556.5=1.997x10^6 \text{ Btu/h}$$

Substituting these values in million Btu's:

$$-4.772 + Q_B = -2.348 - 2.476 + 1.997$$
$$Q_B = 1.945 \text{ million Btu/h}$$

Answer: B

5-2

This is an application of the Rayleigh equation for differential distillation.

$$\ln\left(\frac{L_1}{L_2}\right) = \int_{x_{i2}}^{x_{i1}} \frac{dx_i}{y_i - x_i} = \int_{x_i=0.00005}^{x_i=0.003174} \frac{dx_i}{13x_i - x_i} = \ln\left(\frac{0.003174}{0.00005}\right)^{\frac{1}{12}} = \ln(1.41325)$$

Therefore, L_1/L_2 = 1.41325 or, $L_2 = L_1/1.41325$ = 2287/1.41325 = 1618.3 lbmoles.
Moles to distill = 2287 - 1618.3 = 668.7 lbmoles(**Answer a**). Distillation time = 668.7/200 = 3.35 hrs(**Answer b**).

5-8

The average ammonia concentration in the distillate may be obtained by material balance.
$1618.3(0.00005) + 668.7(x_D) = 2287(0.003174)$ Or, $x_D = 0.01073$ mole fraction(**Answer c**).

The vertical distance for an actual stage = 0.7(10) = 7 mm(**Answer: d**).

5-3

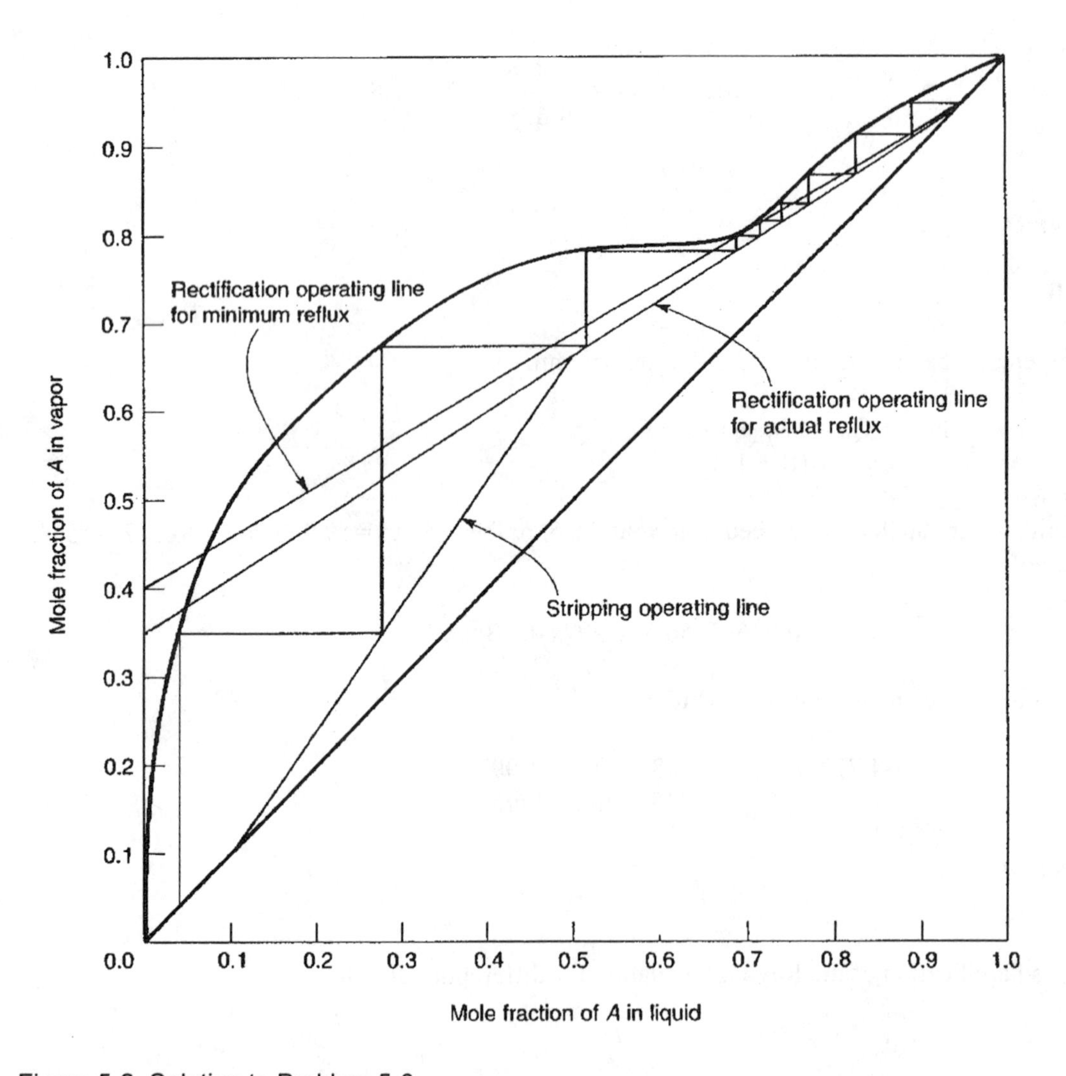

Figure 5-2. Solution to Problem 5-3.

From the intercept of the operating line at minimum reflux,

$$\frac{x_D}{R_m + 1} = 0.4$$

Since $x_D = 0.95$, $R_m = 1.375$, and therefore $R_{op} = 1.375 \times 1.25 = 1.719$. The intercept of the operating line in rectification section is :

$$\frac{x_D}{R_{op} + 1} = \frac{0.95}{1.719 + 1} = 0.349$$

With this intercept, the rectification operating line is drawn. Next, the q-line is drawn. Locating the bottoms composition of 0.1 of light key on the 45° line, the stripping operating line is drawn. The number of theoretical stages is counted to be 9[**Answer (a)**].

The determination of actual number of stages requires the estimation of plate efficiency. Since the viscosity of the liquid is given, we will use O'Connel plot. First, we determine average relative volatility, α_{ij} (I=light key, j=heavy key).

$$\alpha_{ij} = \frac{\left[\frac{y_i}{x_i}\right]}{\left[\frac{y_j}{x_j}\right]} = \frac{y_i x_j}{x_i y_j} = \left(\frac{y_i}{1 - y_i}\right)\left(\frac{1 - x_i}{x_i}\right)$$

Location in column	x_i	y_i	α_{ij}
Top	0.95	0.975	2.053
Middle	0.50	0.78	3.545
Bottom	0.1	0.495	8.822

$$(\alpha_{ij})_{av} = (\alpha_T \alpha_M \alpha_B)^{\frac{1}{3}} = 4.004$$

$(\alpha_{ij})_{av}\, \mu_l = 4.004 \text{x} 0.5 = 2.002$
From Perry [5th or 6th edition, p18-14], plate efficiency ≅ 0.41
Therefore, number of actual plates = 9/0.41 ≈ 22.**[Answer:(b)]**

To estimate the reboiler duty, we need the condenser duty.

5-10

$$Q_c = D(R_{op} + 1)\lambda = \left[\frac{F(x_F - x_B)}{x_D - x_B}\right](R_{op} + 1)\lambda$$

$$= \frac{1(0.5 - 0.1)}{0.95 - 0.1}(1.719 + 1)11000 = 14{,}075 Btu/h$$

Next calculate the number of moles of distillate and bottoms per mole of feed.

$$D = \frac{F(x_F - x_B)}{x_D - x_B} = \frac{1(0.5 - 0.1)}{0.95 - 0.1} = 0.471$$

$$B = \frac{F(x_D - x_F)}{x_D - x_B} = \frac{1(0.95 - 0.5)}{0.95 - 0.1} = 0.529$$

Taking the feed temperature as the datum temperature, an energy balance around the whole column gives:

$$Q_B = DC_{PD}(t_D - t_F) + BC_{PB}(t_B - t_F) + Q_C$$
$$= 0.471(50)(150 - 155) + 0.529(51)(160 - 155) + 14{,}075$$
$$= 14092.1 \ Btu/h$$

Answer:(c).

5-4

The equilibrium line in terms of α, and the q-line are given by:

The equilibrium line:

$$y_i^* = \frac{\alpha x_i^*}{1 + (\alpha - 1)x_i^*}$$

The q-line:

$$y_i = \frac{q_i}{q_i - 1}x_i - \frac{x_{Fi}}{q_i - 1}$$

The intersection of the q-line and the equilibrium line gives the coordinates of the pinch point required to determine the minimum reflux. Equating y_i from the above equations, one gets:

$$q_i(\alpha - 1)(x_i^*)^2 + [q_i - \alpha(q_i - 1) - (\alpha - 1)x_{F_i}]x_i^* - x_{F_i} = 0$$

Substituting $q_1 = 1$, $\alpha = 2.1$, and solving , one gets $x_1^* = 0.65$, which is the same as the composition of the first feed. Not surprising, because the feed is at the bubble point!

$$y_1^* = \frac{(2.1)(0.65)}{1 + (2.1 - 1)(0.65)} = 0.796$$

These are the coordinates of the first pinch point at the intersection of the first q-line with the equilibrium curve. The minimum reflux corresponding to this point may be given by:

$$\left(\frac{L_I}{D}\right)_{\min,1} = \frac{x_D - y_1^*}{y_1^* - x_1^*} = \frac{0.972 - 0.796}{0.796 - 0.65} = 1.205$$

The second pinch point goes through the intersection of the second q-line with the equilibrium curve. Since $q_2 = 1$, we know from the experience of the first solution, $x_2^* = x_{F2} = 0.25$.

$$y_2^* = \frac{(2.1)(0.25)}{1 + (2.1 - 1)(2.5)} = 0.412$$

The minmum reflux required by the second pinch point may be given by:

$$\left(\frac{L_I}{D}\right)_{\min,2} = \frac{\left[(1 - q_1)\frac{F_1}{D} - 1\right] y_2^* + q_1\left(\frac{F_1}{D}\right) x_2^* + x_D - \left(\frac{F_1}{D}\right) x_{F_1}}{y_2^* - x_2^*}$$

All parameters of the right hand side of the equation except D are known. D is obtained by a material balance.

5-12

$$D = \frac{F_1 x_{F1} + F_2 x_{F2} - (F_1 + F_2) x_B}{x_D - x_B} = \frac{105x0.65 + 55x0.25 - 160x0.022}{0.972 - 0.022} = 82.6$$

Substituting all numerical values, one gets:

$$\left(\frac{L_I}{D}\right)_{min,2} = 0.318$$

The controlling minimum reflux is the higher of the two, and is 1.205(**Answer**).

ABSORPTION

6-1.

Size the diameter of a countercurrent absorption column with the following available information.

Vapor into the column is air at 90°F and 15 psia and at a flow rate of 3600 ACFM containing ammonia at a concentration of 250 ppm(v/v).
The concentartion of ammonia is to be reduced to 0.15 ppm, using water at 90°F as once through operation.

The equilibrium relation of ammonia-air-water for dilute solutions is given by:

$\ln(y/x) = -4425/T + 14.8374$
where:
y = mole fraction of ammonia in air
x = mole fraction of ammonia in water
T = temperature in °K

Two sizes of pall rings(plastic) are available:

Size:	1"	2"
Packing factor, F_p	52	25
Specific surface, A_p, ft^2/ft^3	63	31

6-2.

A packed column with 4' internal diameter filled with 2" pall ring will be used to scrub an air stream at 16000 lb/h, 90°F, and 15 psia containing 300 ppm(v/v) of ammonia to 0.2 ppm using 36000 lb/h of water at 90 °F countercurrently. The equilibrium relationship of ammonia-air-water for dilute solutions may be given by:

$y = 1.4098x$
where:
y = mole fraction of ammonia in air
x = mole fraction of ammonia in liquid

The overall mass transfer coefficient is:
K_Ga = 13.65 lb.mol/h.cuft.atm at 4 gpm/ft^2 of liquid rate and 3.5 ft/s superficial velocity of gas.
Assume that K_Ga is proportional to $V^{0.88}xL^{0.09}$. Where V = superficial gas velocity, ft/s and L = liquid rate, gpm/ft^2.

6-2

6-2.1 The density of entering gas(lb/cu.ft) is close to:
(A) 1 (B) 28.95 (C) 0.1 (D) 0.3

6-2.2 The superficial gas velocity(ft/s) is approximately:
(A) 10 (B) 1 (C) 3 (D) 5

6-2.3 The liquid rate in gpm/ft^2 cross section is close to:
(A) 20 (B) 1 (C) 10 (D) 6

6-2.4 If the average gas density is 0.073 lb/ft^3 and liquid rate(gpm/ft^2 cross-section) is 5.73, the corrected overall mass transfer coefficient K_Ga, lb.mol/h.cuft.atm., is close to:
(A) 40 (B) 10 (C) 19 (D) 5

6-2.5 If K_Ga of the system is 18.76 lb.mol/h.cuft.atm , then the overall height of transfer unit, H_{OG}, ft/transfer unit is approximately:
(A) 4 (B) 2 (C) 1 (D) 10

6-2.6 The slope of the operating line is close to:
(A) 4 (B) 0.25 (C) 2 (D) 0.44

6-2.7 The slope of equilibrium line,on mol fraction basis, is :
(A) 1.4098 (B) 0.7093 (C) y/x (D) 1

6-2.8 If the slope of the operating line = 3.619 and the slope of the equilibrium line is 1.4098, then the number of transfer unit is close to:
(A) 18 (B) 3 (C) 20 (D) 11

6-2.9 If H_{OG} is 2.3 ft, and N_{OG} is 11.17, then the height of packing(ft) is close to:
(A) 5 (B) 26 (C) 13 (D) 10

6-2.10 If the slope of the operating line is 3.619 and N_{OG} = 11.17 then the estimated number of theoretical plates to perform the same duty in a column is:
(A) 7 (B) 11 (C) 22 (D) 5

SOLUTIONS

6-1.

(The solution follows steps on pp 203 - 206 of CELR)
Step 1. Estimate the gas flow rate,G, in lb/h

Since the concentration of ammonia is small, the contribution of solute is ignored.
Density of air ρ_{air} = PM/RT = 15x28.95/(10.73x(460+90)) = 0.073 lb/ft^3.

Gas flow in = 3600 ft^3/m x 0.073 lb/ft^3
= 262.8 lb/m = 15768 lb/h
Step 2. Translate the given equilibrium relationship into the form y = mx, and find out the value of m.

ln(y/x) = - 4425/T + 14.8374
Operating temperature = 90 °F = 305.3 K
ln(y/x) = -4425/305.3 + 14.8374
Hence, y = 1.4098x
Therefore, m = 1.4098

Step 3. Select a packing size.

From the given gas flow rate and the guideline in Table 6-1 of CELR, select 2" packing from the given data. F_p = 25.

Step 4. Select unit pressure drop.

Using the guideline in Table 6-2 of CELR, select 0.25" wc/ft of packing.

Step 5. Estimate the water flow rate.

Water flow rate, lb/h = 1.6m(gas flow rate, lb/h)
= 1.6x1.4098x15768 = 35567.6 lb/h [See p196 step 2 of CELR).

Step 6. Calculate the x-parameter of Figure 6-4 of CELR.

Density of water at 90 °F = 62.12 lb/ft^3.
$(L/G)(\rho_G/\rho_L)^{0.5}$ = $(35567.6/15768)(0.073/62.12)^{0.5}$
= 0.077
Looking at Figure 6-4 of CELR, this is within the range of the x-parameter.

Step 7. Find out the y-parameter of Figure 6-4 of CELR, and calculate the column diameter.

Corresponding to the unit pressure drop of 0.25 inH_2O/ft and x-parameter of 0.077, read y-parameter as 0.73

6-4

$$\frac{CG'^2 F_p v^{0.1}}{\rho_G(\rho_L - \rho_G)} = 0.73$$

Where:
C = 1 for British units
F_p = 25 [step 3]
v = 1 for water

$\rho_G = 0.073$ lb/ft^3
$\rho_L = 62.12$ lb/ft^3
Substituting these values:

$$G' = \left(\frac{0.73 \times 0.073 \times (62.12 - 0.073)}{25 \times 1}\right)^{0.5} = 0.3637 \text{ lb/ft}^2\text{.s}$$

Column diameter = $0.0188(G_T/G')^{0.5} = 0.0188(15768/0.3637)^{0.5} = 3.915$ ft

[Note: G_T = 15768 from step 1]

Use a 4 ft diameter column.

Step 8. Finally check the wetting of the packing.

Wetting rate of packing
= gpm of water/cross section of column.
= $(35567.6/500)/(\pi/4)x16 = 5.66$ gpm/ft^2.
From Table 6-4 of CELR, the minimum wetting rate of the packing = $0.106A_p = 0.106x31 = 3.3$ gpm/ft^2.

Therefore, the 4 ft diameter column is satisfactory.

6-2.

6-2.1 Density of gas = 15x28.95/(10.73x(460+90) = 0.073 lb/ft^3. **Answer: C**

6-2.2 Superficial velocity =1600x4/(0.073x3600xπx16) = 4.84 ft/s. **Answer: D**

6-2.3 Liquid rate = 36000x4/(500xπx16) = 5.73 gpm/ft^2. **Answer: D**

6-2.4 Correction factor =$(4.84/3.5)^{0.88}$x$(5.73/4)^{0.09}$ = 1.374.
Corrected K_Ga =13.65x1.374=18.76lb.mol/h.cuft.atm. **Answer: C**

6-2.5 $H_{OG} = \dfrac{G}{K_g a P_t (1-y)_{lm}}$

G = molar gas mass velocity. = 16000x4/(πx16x28.95) = 43.98 lb.mol/h.ft^2
[solute contribution is ignored.]

P_t = 15/14.7 = 1.0204 atm. Since solute concentration is low, $(1-y)_{lm}$ may be assumed 1.
H_{OG} = 43.98/(18.76x1.0204x1) = 2.3 ft/transfer unit. **Answer: B**

6-2.6 Since ammonia concentration in air is low, solute free air mass velocity, G_s, may be computed as G_s = 16000x4/πx16x28.95 = 43.98 lb.mol/h.ft^2. Similarly, solute free water mass velocity, L_s, may be calculated as L_s = 36000x4/πx16x18 = 159.155 lb.mol/h.ft^2.
Slope of the operating line = L_s/G_s = 3.619. **Answer: A**

6-2.7 Slope of the equibrilium line = 1.4098. **Answer: A**

6-2.8 From the given data, the operating line and the equilibrium line are not parallel. So, the equation (6-16) of CELR may be used to calculate the number of transfer units.

$$N_{OG} = \frac{1}{1 - mG/L} \ln\left(\frac{(1 - mG/L)(y_1 - mx_2)}{y_2 - mx_2} + mG/L\right)$$

Where:
N_{OG} = number of transfer units. m = slope of equilibrium line = 1.4098.
G/L = ~ G_s/L_s = 1/3.619 = 0.2763
y_1 = concentration of ammonia in inlet vapor, mole fraction = 0.0003, given.
y_2 = concentration of ammonia in outlet vapor, mole fraction =0.0000002, given.
x_2 = concentration of ammonia in inlet liquid, mole fraction. = 0, assumed.

Substituting these values,

N_{OG} = 11.17 **Answer: D**

6-2.9 Height of packing = H_{OG} x N_{OG} = 2.3x11.17 = 25.69 ft
Answer: B

6-2.10 From equation (6-24) of (CELR):

$N_P = N_{OG}$x(A-1)/lnA
where:
N_P = number of theoretical plates.
N_{OG} = 11.17, given.
G/L = 1/(L/G) = 1/3.619 = 0.2763 A = mG/L = 1.4098x0.2763 = 0.3895

6-6

Substituting these values:

$N_P = 11.17 \times (0.3895-1)/\ln(0.3895)$
$= 7.23$ **Answer: A**

LEACHING

7-1: 100 kg of a solid containing 40 % solute A and 60 % inerts B is treated with a solvent C in a single stage single contact extraction. The extracted solids are then screw-pressed. The pressed solid contains 1.2 kg solution per kg of inerts. Entrainment of solids in the extract may be neglected.

(a) Calculate the solvent amount that needs to be used and the mass fraction of the solute in the extract if 90 % of the solute in the feed is to be recovered in a single stage single contact.

(b) What will be the recovery of the solute if the extraction is carried out in two crosscurrent stages if 50 % of the solvent amount in (a) is used in each stage.

(c) Show the calculations of part b on a right-triangular diagram.

7-2: Oil is to be extracted from meal using benzene as a solvent. The extraction unit is to treat 1000 kg/h of meal (oil-free solid). The feed meal contains 400 kg of oil and 25 kg of benzene. The wash benzene contains 10 kg of oil dissolved in 655 kg of benzene. The discharge solids must not contain more than 60 kg of oil. The solution retention data are given in the following table.

Concentration (kg of oil/kg solution)	0	0.1	0.2	0.3	0.4	0.5	0.6	0.7
Solution retained, kg /kg of inert solid	0.5	0.505	0.515	0.53	0.55	0.571	0.595	0.62
kg of solute/kg of inerts	0.0	0.0505	0.103	0.159	0.22	0.2855	0.357	0.434

(a) Compute and plot a Ponchon-Savarit diagram for this extraction system.
(b) Estimate the following for multistage countercurrent extraction of oil:

1. composition of the strong solution
2. the weight of solution leaving with the extracted meal.
3. amount of the strong solution.
4 number of theoretical stages required.

7-3: 100 tons/h of fresh beets (composition: pulp 40 %, sugar 12 % and water 48%) are to be treated in a countercurrent battery of cells with water containing no sugar. If 97 % of the sugar is to be recovered with at least 15 % sugar in the extract phase, determine the number of cells required if each ton of dry pulp retains 3.5 tons of water. Do not use graphical construction.

7-4: 1000 kg of a natural material containing 30 % oil and 70 % inert pulp is to be treated with a solvent C to extract the oil. The extracted solid will be pressed so that the pulp retains 1 kg of solution per kg of pulp. Calculate the solute and inert free solvent that needs to be used and the solute concentration in the extract phase if 90 % of the oil is to be recovered.

7-5: The underflow compositions for the leaching of a solute from an ore containing 20 % solubles, 15 % water, and rest inerts are plotted in the triangular diagram (Figure 7-5). The solubles are to be extracted in a countercurrent multistage system. The feed rate to the system is to be 20 t/h. A solute recovery of 90 % is desired.

(a) What is the quantity of the inerts in the underflow from each stage?
(b) How much solute will be discarded with the underflow discharge from the system?
(c) How much water will be present in the underflow discharge from the system?
(d) What is the minimum water-to-inerts ratio that can be used to effect the same separation?
(e) What are the coordinates of the difference point?

7-6 :

7-6.1: 10 kg of a solid containing 45 % of a soluble material were treated with 15 kg of a solvent containing the same solute at 2% concentration in a vessel under constant agitation. After a long time the solution and the solid were separated by pressing. The solid analyzed 0.8 kg of solvent per kg of inert solid. The extract quantity obtained in kg was therefore

a) 14.8 b) 15.1 c) 13.7 d) 14.8

7-6.2: In a determination of the solution retention data, the mass fraction of the solute in the extract was determined to be 0.6. The corresponding underflow analysis showed a retention of 0.6 kg solvent per kg of the inert solid. The mass fraction of solute in the underflow is most likely

a) 0.6 b) 0.225 c) 0.36 d) 0.375

7-6.3: In the continuous extraction of sugar (A) from beets (see figure below) the inert pulp (I) retains 3 tons of solution per ton of inerts in the underflow from stage to stage. 100 tons/h of fresh beets are to be processed per hour. 97 % of A is to be recovered in the extract.

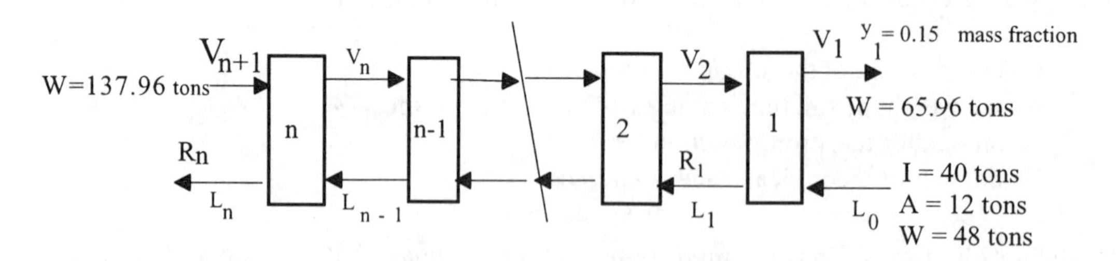

7-6.3.1 The mass fraction of solute A in V_2 is most likely

a) 0.15 b) 0.12 c) 0.1282 d) 0.1

7-6.3.2 The number of cells in the extraction system is most likely to be

a) 15 b) 17 c) 15.5 d) 16

SOLUTION TO PROBLEMS

7-1:

(a) solute in feed = 100 x 0.4 = 40 kg
inert solids in feed = 100 - 40 = 60 kg
solution retained by the inert solid = 1.2 x 60 = 72 kg
solute in extract = 0.9 x 40 = 36 kg
solute in inerts = 40 - 36 = 4 kg
$R_1 = 60\,(1.2) + 60 = 132$ kg $L_1 = 72$ kg
By solute balance, $40 + 0 = V_1\,y_1 + 72\,x_1$
But $x_1 = y_1$ Therefore $V_1\,y_1 + 72y_1 = 40$
$y_1 = (4/72) = 0.0556$ mass fraction
substituting value of y_1, $V_1 = 648$ kg
By overall balance, fresh solvent = $V_2 = V_1 + R_1 - R_0$
$= 648 + 132 - 100 = 680$ kg
Fresh solvent = 680 kg and solute mass fraction in extract = 0.0556

(b) solvent to be used in each stage = 680/2 =340 kg
Material balance on first stage:
solution in pressed solids = 1.2(60) = 72 kg
By overall balance, extract $V_1 = 100 + 340 - 1132 = 308$ kg
By solvent balance, $340 = 72(1 - y_1) + 308(1 - y_1)$
from which $y_1 = 0.105263$
solute in extract = 0.105263 (308) = 32.42 kg
solute in solids = 40 - 32.42 = 7.58 kg

Material balance on second stage.:
By overall balance extract amount = 132 + 340 - 132 = 340 kg
Solute in solution = 7.58 kg total
total solution = 340 + 72 = 412 kg
mass fraction solute in solution = (7.58/412) = 0.0184
solute in extract from second parallel stage = 0.0184(340) = 6.26 kg
solute recovered in two stages = 32.42 + 6.26 = 38.68 kg
% recovery = (38.68/40)(100) = 96.7 %
Therefore the recovery has increased from 90 to 96.7 %

(c) The construction is shown in figure 7- 1.

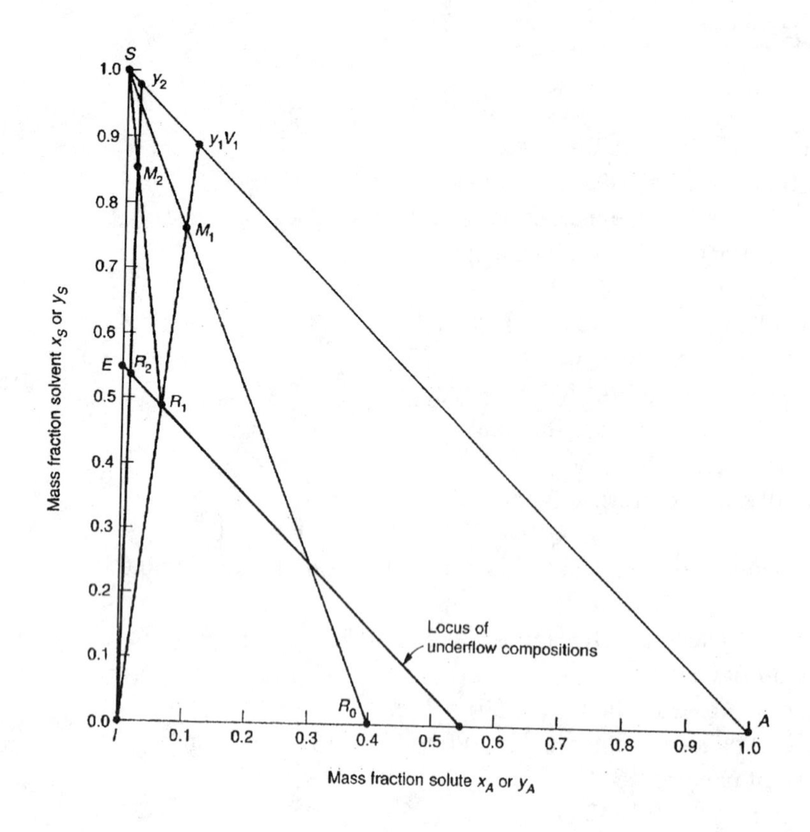

Figure 7-1: Solution of two stage cross-current extraction (Problem 7-1)

(Note: The following explanation is not part of answer. Construction of the diagram is adequate. The explanation is given as a review.)

Solution retention data given in a and b above imply the inerts in the underflow retain 1.2 kg/kg inerts and this value is constant. Locus of underflow compositions on triangular diagram is a straight line parallel to hypotenuse ($x_I = 0$). and intersects the vertical ordinate in $x_S = \frac{K}{K+1} = \frac{1.2}{1.2+1} = 0.5455$ and $x_A = 0$. This line is drawn as EG on the diagram. point R_0 representing feed composition to first stage and solute-free solvent , and point S are joined. The addition point M_1 is located on line $\overline{R_0S}$ by knowing amounts of feed and solvent to be used. A tie line is drawn passing through the point M_1 and vertex I to cut EG (underflow composition locus) in R_1 and hypotenuse in y_1 . Knowing fresh solvent to be used in second stage and R_1 and their compositions, point M_2 , the addition point is located on the line $\overline{R_1S}$. A second tie line is drawn from origin to pass through M_2 and to intersect the underflow locus in R_2 and the hypotenuse in y_2.. The point R_2 gives the composition of the solids discharged from the last stage and the point y_2 gives the composition of final extract.

7-2:

(a) Calculate the values of N = kg of inerts/kg of solution for various concentrations of overflow solutions

kg of oil/kg of solution	0	0.1	0.2	0.3	0.4	0.5	0.6	0.7
N= kg inerts/kg of solution	2.0	1.98	1.95	1.87	1.818	1.751	1.68	1.613

Plot N vs x_A to get the locus of the underflow curve. **(Figure 7-2)**

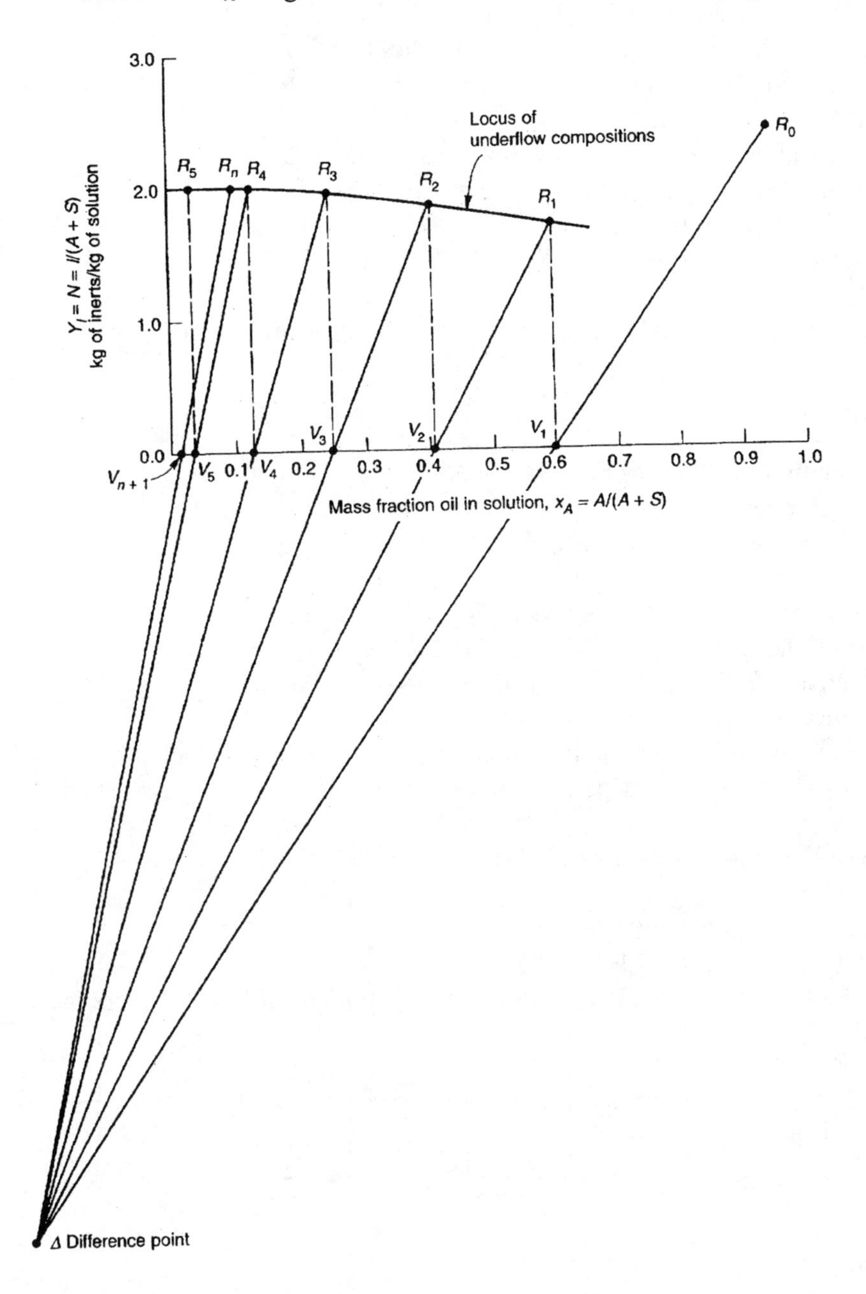

Figure 7-2: Solution of Problem 7-2 on Ponchon-Savarit diagram

(b) 1. composition of the strong meal

For R_0, $N = 1000/(400 + 25) = 2.353$ and $x_A = 400/425 = 0.941$

For V_{n+1}, $N = 0$ and $x_A = 10/(10 + 655) = 0.015$

Discharged solids :

inerts in solids = 1000 kg

solute in solids = 60 kg

Solute/inerts ratio = 60/1000 =0.06

By interpolation from the table, mass fraction of solvent in solution

= (0.0095/0.0525(0.1) + 0.1 = 0.118

Solvent in R_n = [(1-0.118)/0.118] (60) = 448 kg

By solvent balance , solvent in extract = 655 + 25 - 448 = 232 kg

Oil in extract (overflow) = 350 kg

mass fraction of oil in strong solution = (350/(350 + 232) = 0.601

(2) Weight of solution leaving with the extracted meal

= 60 + 448 = 508 kg /h

(3) Amount of strong solution. = 350 + 232 = 582 kg/h

(4) Number of theoretical stages required = 4.3 (approx. from graph)

7-3:

1. **Basis of calculation 100 tons of fresh beets**

Overall balance:

sugar to be recovered = 0.97 x 12 = 11.64 tons

sugar in discarded pulp = 12 - 11.64 = 0.36 tons

water in extract phase = (0.85/0.15)(11.64) = 65.96 tons

by water balance, $V_{n+1} + 48 = 40 \times 3.5 + 65.96$ i.e

Then V_{n+1} (fresh water) = 140 + 65.96 - 48 = 157.96 tons

water in discarded pulp = 3.5 x 40 = 140 tons

underflow to and overflow from the first stage are not the same as in the other stages i.e. L_0 $L_1 = L_2$ ···. Therefore a separate material balance is required on first stage.

2. **Material balance over first stage:**

$V_2 + R_0 = V_1 + R_1$

water in R_1 = 40 x 3.5 = 140 tons

sugar in R_1 = (0.15 /0.85)(140) = 24.7 tons

R_1 = 40 + 140 + 24.7 = 204.7 tons (since concentrations of L_1 and V_1 are the same)

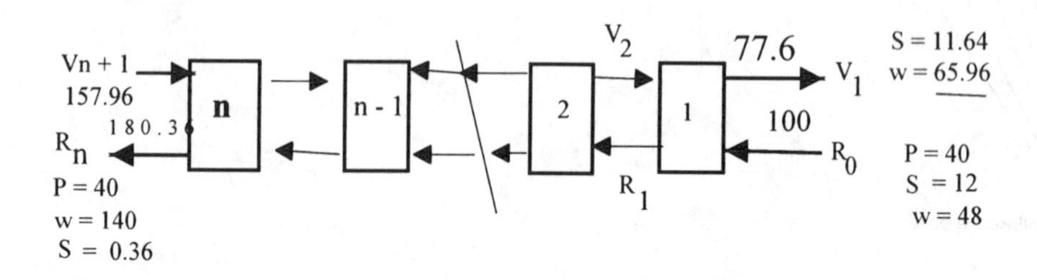

Figure 7-3 : Material balance for multistage extraction in problem 7-3.

$V_2 = V_1 + R_1 - R_0 = 77.6 + 204.7 - 100 = 182.3$ tons
by sugar balance, $V_2 y_2 = 11.64 + 24.7 - 12 = 24.34$ tons
water in $V_2 = 182.3 - 24.34 = 157.96$ tons

3. Calculation of number of stages

use analytical equation of Smith-Baker to determine the number of remaining (n - 1) stages. The following quantities are needed.

$$y'_{n+1} = \frac{0}{157.96} = 0 \quad x'_n = \frac{0.36}{140} = 0.0025 \quad y'_1 = \frac{24.34}{157.96} = 0.1541 \text{ and } x'_1 = \frac{24.7}{140} = 0.176$$

These are ratios of mass of solute to mass of solvent which are required when solvent retained by inerts is constant from stage to stage.

$$n - 1 = \frac{\log\left[\left(y_{n+1} - x_n\right)/\left(y'_1 - x'_1\right)\right]}{\log\left[\left(y'_{n+1} - y'_1\right)/\left(x'_n - x'_1\right)\right]} = \frac{\log[(0 - 0.00257)/(0.1541 - 0.1}{\log[(0 - 0.1541)/(0.00257 - 0.1} = 17.92$$

Then n = 17.92 + 1 = 18.92 or 19 cells since number of cells must be an integral number.

7-4:

Single stage single contact extraction:

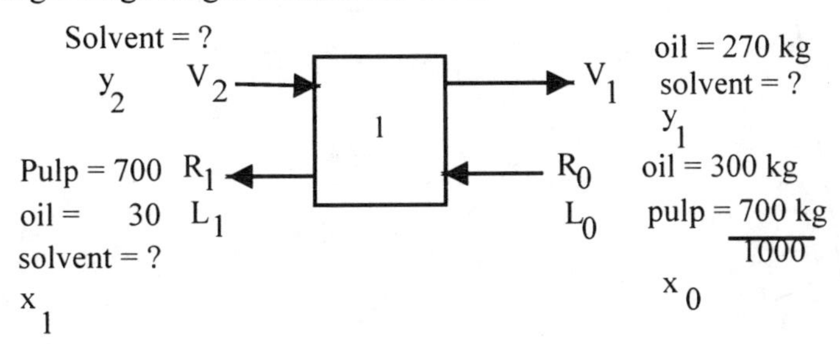

Overall material balance:

$$V_2 + L_0 = V_1 + L_1$$

Solute balance :

$$V_2(y)_2 + R_0(x_A)_0 = V_1 y_1 + R_1(x_A)_1$$

Assume equilibrium is reached. then $y_e = x_e$
$R_0 = 1000$ kg, $L_0 = 300$ kg, $L_1 = 700 + 700 = 1400$ kg
With 90 % recovery, oil in extract phase = 0.9 x 300 = 270 kg
Therefore oil in R_1 = 300 - 270 = 30 kg
Solvent in R_1 = 700 - 30 = 670 kg
Solvent in extract phase = (670/30) (270) = 6030 kg
by solvent balance V_2 = 6030 + 670 - 0 = 6700 kg
The concentration of oil in extract phase = (270/(270 + 6030) = 4.29 %

7-5:

Basis 20 t/h

(a) solute in feed = 20 (0.2) = 4 tons

water in feed = 20 x0.15 = 3 tons

Inerts in feed = 13 tons

Since ideality is to be assumed, the discharged solids from the system = 13 tons. This will be also the inert flow from stage to stage in the underflow.

(b) 90 % recovery. therefore solute in extract = 4(0.9) = 3.6 tons

Then solute in the discharged solids = 4.0 - 3.6 = 0.4 ton

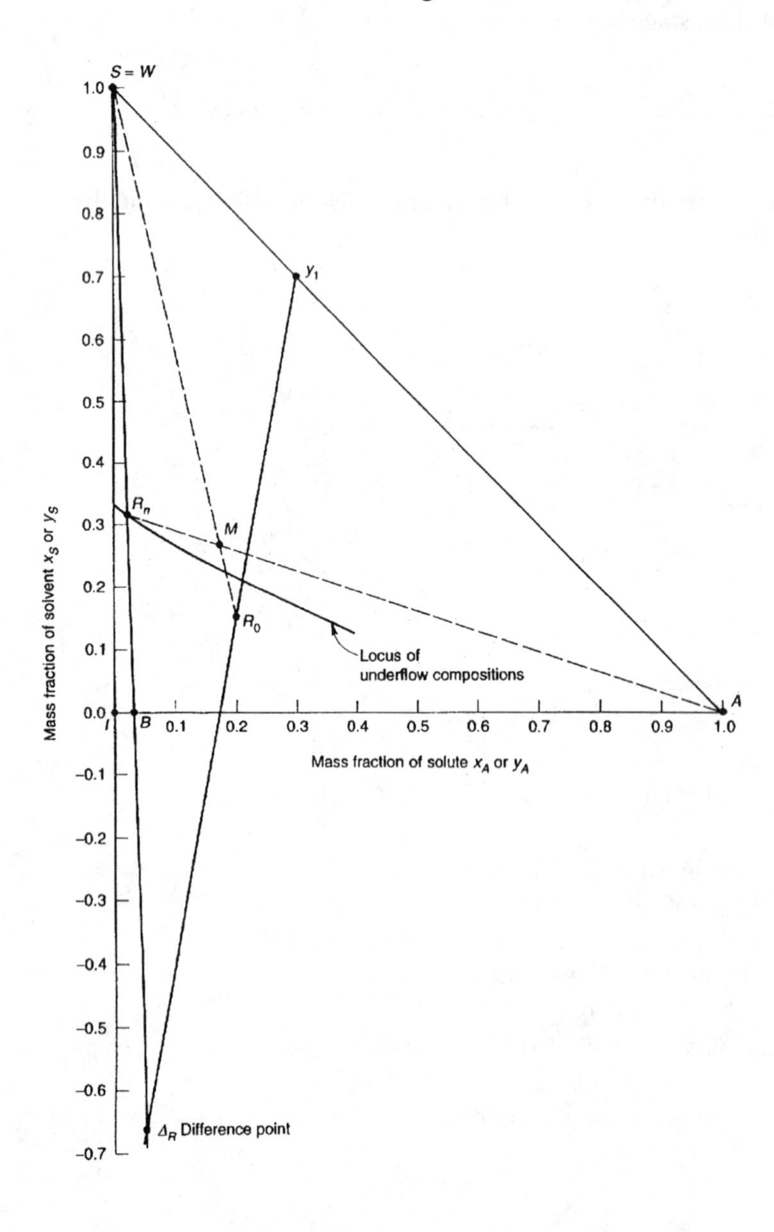

Figure 7-5 : Solution of problem 7-5 on triangular diagram

(c) Mark the point representing R_0, the feed on the diagram

The solute mass fraction in discharged solids on solvent-free basis = 0.4/13.4 = 0.0299

Mark this point on X axis as B and join $\overline{SB}$ to cut underflow line in R_n. This point gives the composition of the discharged solids. From coordinates of R_n,

mass fraction of water = 0.305.

mass fraction of solute = 0.02

Therefore, solvent with the discharged inerts = (1.0 - 0.305 - 0.02) = 0.675

water in the discharged solids = (0.305/0.675)(13) = 5.9 ton

(d) To determine minimum water-to-inerts ratio, join $\overline{R_0S}$ and $\overline{R_nA}$ to intersect each other in M.

$$\text{Minimum water-to-inert ratio} = \frac{\overline{MR_0}}{\overline{My_{n+1}}} = \frac{1.6}{9.6} = 0.1$$

(e) To find coordinates of the difference point, join $\overline{R_0V}$ and $\overline{R_ny_{n+1}}$ and extend them to intersect at the difference point. The coordinates of the difference point are:

$(x_A)_\Delta = +0.05$, $(x_S)_\Delta = -6.6$ read from the diagram.

7-6:

7-6.1

solvent in feed = 0.98x15 = 14.7 kg Solvent in pressed solid = 0.8(5.5) = 4.4 kg

By solvent balance, solvent in extract = 15 x 0.98 + 0 - 4.4 = 10.3 kg

Assume equilibrium is reached, hence $x_e = y_e$

$$\text{solute concentration in solution} = \frac{15\times0.02 + 4.5}{14.7 + 15\times0.02 + 4.5} = 0.2462$$

$$\text{Solute in extract} = \frac{0.2462}{1.0 - 0.2462}(10.3) = 3.36 \text{ kg}$$

Therefore extract solution = 10.3 + 3.36 = 13.66 ≐13.7 kg

Answer c)

7-6.2

Solvent retained per kg of inert solids = 0.6 kg

concentration of solute in the overflow solution = 0.6 mass fraction

Solvent mass fraction in overflow solution = 1- 0.6 = 0.4

underflow containes 0.6 solvent . then solute in underflow = (0.6/0.4)(0.6) 0.9 kg

total underflow per kg of inert solids = 1.0 + 0.6 + 0.9 = 2.5 kg

Therefore mass fraction of solute in the underflow = $x_A = \frac{0.9}{2.5} = 0.36$

Answer c)

7-6.3

7-6.3-1 Take material balance over first stage

quantity of R_1 = 40 + 3(40) = 160 ton/h

A to be extracted = 0.97(12) = 11.64 t/h

Total extract = 11.64 + 65.96 = 77.6 t/h

By total material balance $V_2 = L_1 + V_1 - L_0$ = 120 + 77.6 - 60 = 137.6 t/h

Solute in the underflow from stage 1 = 0.15(120) = 18 t/h

By solute balance, solute in V_2 = 18 +11.64 - 12= 17.64 t/h

Solute concentration in V_2 = 17.64/137.6 = 0.1282 mass fraction

Answer c)

7-6.3-2 Use Smith-Baker equation to get the number of stages. For this first stage has to be evaluated separately because the underflow to this stage is different and the equation cannot be applied if the underflow is not the same. Therefore

$$N-1=\frac{\log\left[\left(y_{n+1}-x_n\right)/\left(y_2-x_1\right)\right]}{\log\left[\left(y_{n+1}-y_1\right)/\left(x_n-x_1\right)\right]}=\frac{\log[(0-0.003)/(0.1282-0.15)]}{\log[(0-0.1282)/(0.003-0.15)]}=15.5$$

Actual number of cells has to be a whole number . The designer will use next whole number. Therefore number of cells in the system = 16

Answer d)

LIQUID-LIQUID EXTRACTION

8-1 :

8-1.1 : The following tie line data were obtained for a ternary system of solute A , extract solvent S and raffinate solvent B at a temperature of 25 °C

	Raffinate layer wt %	Extract layer wt %
A	25.0	42.5
B	67.5	24.6
C	7.5	32.9

The selectivity at this tie line is very nearly

a) 1.7 b) 4.67 c) 0.39 d) 1.0

8-1.2: The distribution coefficient of a solute A between solvents B and S is given by Y = 2.5X where Y = mass of A/mass of S in extract and X = mass of A/mass of B in raffinate. S and B are mutually immiscible. A solution containing 25 % A in B is to be extracted in a single stage contact with a recovery of 80 %. The amount of S in kg required per 100 kg of solution is nearly

a) 100 b) 150 c) 120 d) 200

8-1.3: The distribution coefficient of a solute A between solvents B and S is given by Y =3X where Y and X are mass ratios of A to solvents in extract and raffinate respectively. B and S are mutually insoluble in each other. 100 kg of a 30 % solution of A in B is to be successively treated in two batch contacts with 60 kg of solvent each time. The overall percent recovery of A is very nearly

a) 89.9 b) 92.16 b) 96.5 b) 100

Binodal curve for the system water, acetic acid, and MIBK (Sherwood, Evans & Longo, Ind. Eng. Chem. V31(1939), p1144) is plotted in Figure 8.1a and the distribution diagram is given in Figure 8.1b. 100 lb of a solution of acetic acid in water containing 40 % acetic acid is to be extracted with pure MIBK as the solvent in a single stage extraction. The extraction is isothermal at 25 °C. Answer the following questions by choosing the correct solution.

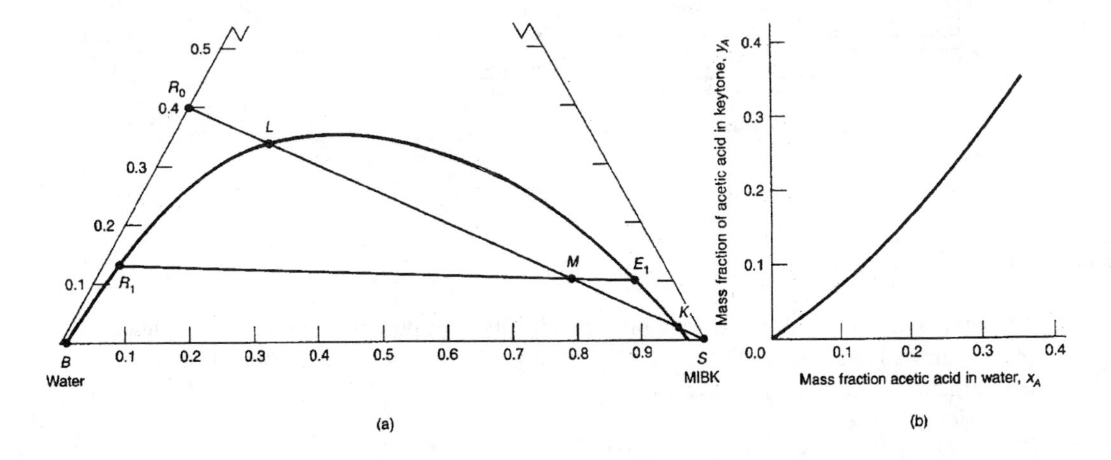

Figure 8.1 a and b

8-1.4 Minimum solvent that can be used is very nearly

a) 33 b) 18.7 c) 25 d) 40

8-1.5. Maximum solvent that can be used is very nearly

a) 871 b) 1000 c) 500 d) no limit

8-1.6 If solvent used is 275 lb, the extract obtained will be very nearly

a) 325 b) 190 c) 350 d) 275

8-1.7 If the amount of the solvent used is 275 lb, percent acetic acid recovery is very nearly

a) 81.25 b) 90 c) 80 d) 98

The binodal and conjugate curves for a ternary system (solute A, solvent S and raffinate solvent B are plotted in Figure 8.2. Some tie lines are also shown .

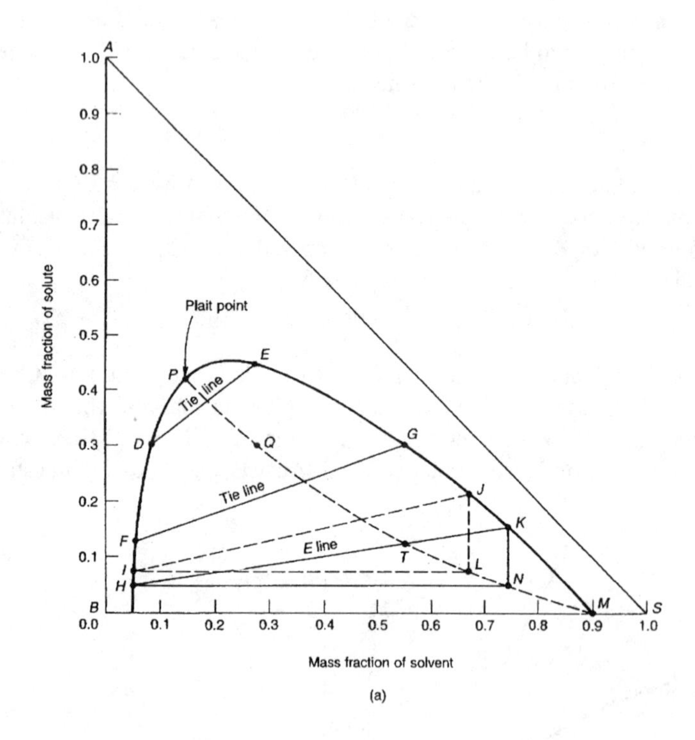

Figure 8.2

8-1.8 100 kg of a mixture of A and B were mixed with 100 kg of pure S and the two conjugate phases allowed to separate. The extract layer analyzed 21 % A and 67.4 % B. The solute mass fraction in the raffinate is very nearly

a) 0 .051 b) 0.10 c) 0.072 d) 0

8-1.9 The amount of raffinate layer in kg is very nearly

a) 80 b) 90 c) 100 d) 56

8-1.10 The wt % concentration of A in the original mixture was very nearly

a) 40 b) 33.15 c) 50 d) 0.55

8-2: The following conjugate phase data (Vriens and Medcalf, Ind Eng. Chem. 45,(1098)1953) are available on the ternary system water (B), pyridine(A), and benzene (S).

Table 8-1: Conjugate phase data for ternary system water-pyridine-benzene are given below.

Data				Calculated	
Raffinate layer		Extract layer			concentration of A
pyridine	Water	pyridine	water	Selectivity	in solvent-free raffinate
0.051	0.947	0.139	0.0075	344.1	0.0511
0.122	0.872	0.27	0.018	107.2	0.1227
0.259	0.716	0.353	0.031	31.48	0. 2656
0.417	0.545	0.384	0.048	10.46	0.4335
0.537	0.363	0.438	0.058	5.1	0.5967

Compute (a) the pyridine selectivities of benzene for various conjugate phases and (b) plot the data against the concentrations of pyridine in raffinate on solvent-free basis.

8-3: The distribution equilibrium for A between an extract solvent S and a raffinate solvent B is given by

$$Y = 2.0\,X$$

where Y = mass of A per unit mass of S, X = mass of A per unit mass of B. The extract and raffinate solvents are completely immiscible in each other at all concentrations of A. From these data, calculate the amount of extract solvent needed per 100 kg solution containing 30 % A in B if 95 % of A is to be removed for each of the following arrangements:

(a) a single -stage contact
(b) three stage batch contact , one third of the solvent will be used in each contact. Calculate amount of A extracted in each contact.
(c) three-stage countercurrent operation

8-4: 5000 lb/h of a solution containing 35 % acetic acid (A-solute) and rest water (B-raffinate solvent) is to be extracted with isopropyl ether (S-solvent) in a multistage countercurrent extraction system. The acid concentration is to be reduced to 2 % in the solvent free raffinate. Determine

(a) minimum solvent to feed ratio and total solvent needed
(b) number of theoretical stages required. if the solvent to be used is 2 times the minimum.

The equilibrium and some tie lines data for the system (Trans. AIChE, **36,** 628(1940), are plotted in Figure 6.

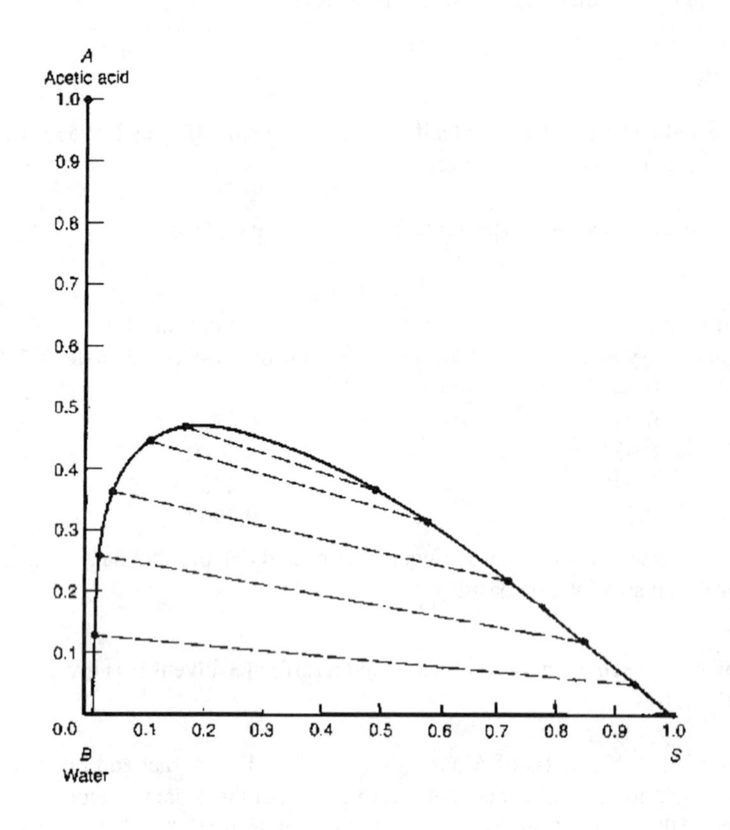

FIGURE 8-3. EQUILIBRIUM AND SOME TIE LINES FOR PROBLEM 8-4.

SOLUTIONS TO PROBLEMS:

8-1:

8-1.1 :

$$\text{Selectivity of S for A} = \frac{y_A/y_B}{x_A/x_B} = \frac{0.425/0.246}{0.25/0.675} = 4.67$$

Answer is b)

8-1.2:

Basis 100 kg of feed.

80 % recovery. Therefore, amount of A in extract = 100(0.25)(0.8) = 20 kg

Amount of A in the raffinate = 25 - 20 = 5 kg

Let S be the amount of solvent in kg per 100 kg of feed

Since extract and raffinate solvents are mutually insoluble ,

amount of B in the raffinate = 75 kg

amount of S in extract = S kg/100 kg of feed

Therefore $\frac{Y}{X} = \frac{20/S}{5/75} = 2.5$

Then $S = \frac{20 \times 75}{2.5 \times 5} = 12$ kg

Answer is c)

8-1.3:

Feed contains 70 kg of B and 30 kg of solute A
Thus S = 60 kg (given) and B = 70 in each stage K = 3 (given) n = 2 stages or contacts
$X_0 = 30/70$ $X_2 = ?$

$$X_2 = \left(\frac{B}{B+KS}\right)^2 X_0 = \left(\frac{70}{70+3\times 60}\right)^2\left(\frac{30}{70}\right) = 0.0336$$

Solute in final raffinate = 0.0336 x 70 = 2.352 kg
Solute recovered in extracts = 30 - 2.352 = 27.648 kg
Then, overall recovery = 92.16 %

Answer is b)

8-1.4:

The points R_0 and solvent E_{n+1} are easily located on the diagram. to start with. Join $\overline{R_0S}$ to intersect the binodal curve in L and K. Point L gives the minimum solvent

$$\text{Minimum solvent ratio} = \frac{S}{R_0} = \frac{\overline{R_0L}}{\overline{LS}} = \frac{1.25}{6.7} = 0.187$$

Amount of minimum solvent = 0.187(100) = 18.7 lb

Answer is b)

8-1.5

The point K gives the condition of maximum solvent

$$\text{Maximum solvent ratio} = \frac{\overline{LS}}{\overline{R_0L}} = \frac{7.625}{0.875} = 8.71$$

Amount of maximum solvent that can be used = 100(8.71) = 871 lb

Answer is a)

8-1.6

Total mixture = 100 + 275 = 375 lb
Coordinates of the addition point are

solute $\left(x_A\right)_M = \frac{40}{375} = 0.10$

Extract solvent $\left(x_S\right)_M = \frac{275}{375} = 0.734$

Raffinate solvent $\left(x_B\right) = 1.0 - 0.734 - 0.107 = 0.159$

This point is located on $\overline{R_0S}$ as point M
By interpolation and with the help of distribution diagram , locate the tie line passing through M. which intersects the binodal curve R_1 and E_1 .
By material balance, $R_1 + E_1$.= 100 +275 = 375 lb

From diagram, $\frac{E_1}{R_1} = \frac{\overline{R_1M_1}}{\overline{M_1E_1}} = \frac{6.37}{31/32.} = 6.575$

E_1 .= 6.575 R_1

Therefore 6.575 R_1 + R_1 = 375

R_1 = 375/7.575 = 49.5 lb E_1 = 375 - 49.5 = 325.5 lb

Answer is a)

8-1.7 From diagram, solute concentration in the extract = 0.1 mass fraction.
Solute in extract = 325.5(0.1) = 32.5 lb
Therefore, recovery = (32.5/40) x 100 = 81.25 %

Answer is a)

8-1.8 Locate given extract layer composition on the binodal curve. as point J. Draw a line from this point and parallel to BA or y axis to meet the conjugate line in point L. Draw a line from point L parallel to X axis to meet the binodal curve in point I. The composition of raffinate is given by point I and is read as: A = 0.072 , S = 0.052 (mass fractions). and S = 0.876 by difference.

Answer is c)

8-1.9 Let R_l and E_l be the amounts of raffinate and extract respectively.
Then overall balance gives $R_l + E_l = 200$ kg
Solvent balance gives $0.052R_l + 0.674\,E_l = 100$ kg
Solving these two equations, $E_l = 144$ kg and $R_l = 56$ kg

Answer is d)

8-1.10 Solute amount = $0.052\,R_l + 0.21E_l = 0.052(56) + 0.21(144) = 33.15$ kg
Raffinate solvent = 100 - 33.15 = 66.85 kg.
A in original mixture = 33.15 % by weight

Answer is b)

8-2:

(a) The selectivity of benzene (Solvent) is given by $\beta = \dfrac{y_A/y_B}{x_A/x_B}$

Substituting the conjugate phase data for one point, $\beta = \dfrac{0.139/0.0075}{0.051/0.947} = 344.$

Corresponding mass fraction of pyridine
in solvent free raffinate = $\dfrac{0.051}{0.051 + 0.947} = 0.051$

(b) In a similar manner, selectivities and mass fractions of solute are calculated and tabulated in the table below as calculated data. The calculated selectivities are plotted in Figure 4 against the mass fractions of pyridine on solvent free basis as required in part b.

Calculated mass fraction of A in solvent-free raffinate	0.0511	0.1227	0. 2656	0.4335	0.5967
Selectivity	344.1	107.2	31.48	10.46	5.1

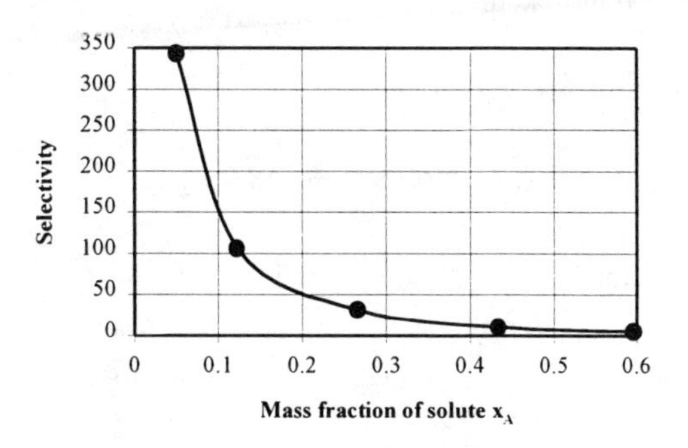

Figure 8-4. Plot of selectivity vs mass fraction of solute for problem 8-2.

8-3:

(a) By solute (A) balance over a single stage,

$$BX_0 + SY_0 = BX_1 + SY_1$$

since fresh extract solvent is free of solute A, $Y_0 = 0$
Basis :100 kg of solution of A in B (raffinate solvent)
$X_0 = 30/70 = 0.42857$
95 % recovery, A in extract = 0.95(30) = 28.5 kg
A in final raffinate = 30 - 28.5 = 1.5 kg
$X_1 = 1.5/70 = 0.02143$ $\quad$ $Y_1 = 2 X_1 = 2(0.02143) = 0.042857$ (by equilibrium)
Then S = B/0.1052639 = 70/0.1052639 = 665 kg

(b) When the extract and raffinate solvent amounts are same in each contact and K is constant, the following relationship applies

$$X_n = \left(\frac{b}{b+KS}\right)^n X_0$$

In this case, n = 3, b = 70, K = 2,

$$X_0 = 30/70 = 0.42857,$$
and $X_n = X_3 = 1.5/70 = 0.0214287$

substituting in the previous equation relating n, X_n and X_0,

$$\frac{X_3}{X_0} = \left(\frac{70}{70+2S}\right)^3 = \frac{0.021428}{0.42857} = 0.05$$

or
$$\frac{70}{70+2S} = (0.05)^{1/3} = 0.3684$$

from which, S = 60 kg in each stage.
Total solvent in three stages = 60 x 3 = 180 kg.

Amount of A removed in first contact:

If x_1 is amount removed in first contact, $X_1 = (30 - x_1)/70 \quad Y_1 = x/60$

By equilibrium relationship, $Y_1 = 2X_1$

Therefore, $x/60 = 2[(30 - x)/70]$

solving for x_1, $$x_1 = \frac{30}{1.583333} = 18.95 \text{ kg}$$

in a similar manner, $$x_2 = \frac{30 - 18.95}{1.583333} = 6.98 \text{ kg}$$

and $$x_3 = \frac{30 - 18.95 - 6.98}{1.583333} = 2.5 \text{ kg}$$

(c) Three-stage countercurrent operation

An overall solute balance gives

$$SY_{n+1} + BX_0 = SY_1 + BX_n$$

or $$\frac{B}{S} = \frac{Y_1 - Y_{n+1}}{X_0 - X_1}$$

where B = raffinate solvent, S = extract solvent

Y = solute in extract, mass A/unit mass of extract solvent

X = solute in raffinate solvent , mass A/unit mass of raffinate solvent

The above equation represents the operating line and is a straight line.

Since the amount of solvent is not known, the final solute concentration in extract is not known.

Therefore a trial solution is required . Assume a value for the B/S ratio. The problem can be solved graphically more easily. Plot the equilibrium curve, on XY diagram and locate the points X_0 and X_1 on it. Draw from point X_1 an operating line with an assumed slope and step off the stages. If the number of stages does not match 3 exactly, readjustment is necessary. This is done in Figure 8-5.

Figure 8-5. Solution of problem 8-3c.

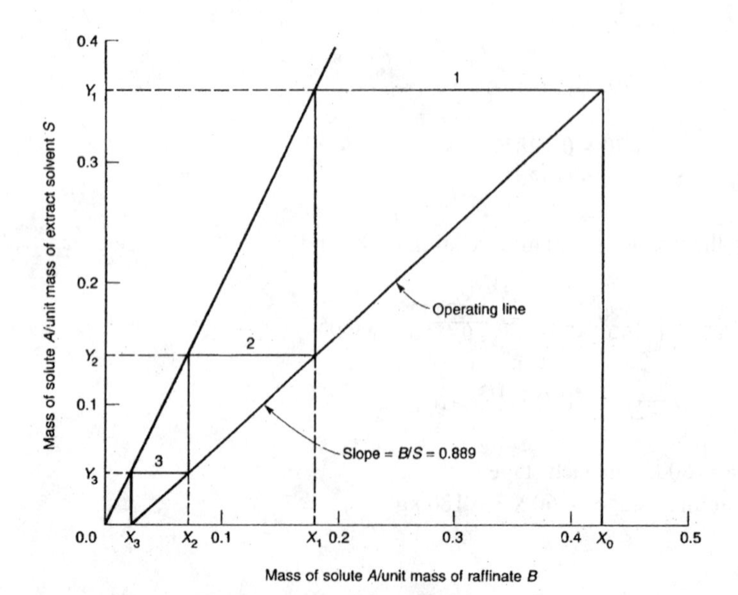

Slope of the operating line for three stages is 0.889 .

Therefore, amount of solvent for 3 stage countercurrent operation is

$$S = \frac{B}{0.889} = \frac{70}{0.889} = 78.74 \text{ kg.}$$

Final extract concentration :

28.5/(28.5 + 78.74) x 100 = 26. 6 wt % .

8-4:

The equilibrium diagram (Figure 6a) provided will be used to solve the problem.
Composition of R'_n on solvent -free basis is 0.02 mass fraction
Join R'_n and S . This line cuts the binodal curve in R_n
The point R_n provides the composition of the final raffinate. $x_A = 0.02$, $x_S = 0.014$, $x_B = 0.966$
Join R_0 and S . By interpolation and by trial draw the tie line that passes through R_0 .

Figure 8-6a. Equilibrium diagram of acetic acid and water solutions

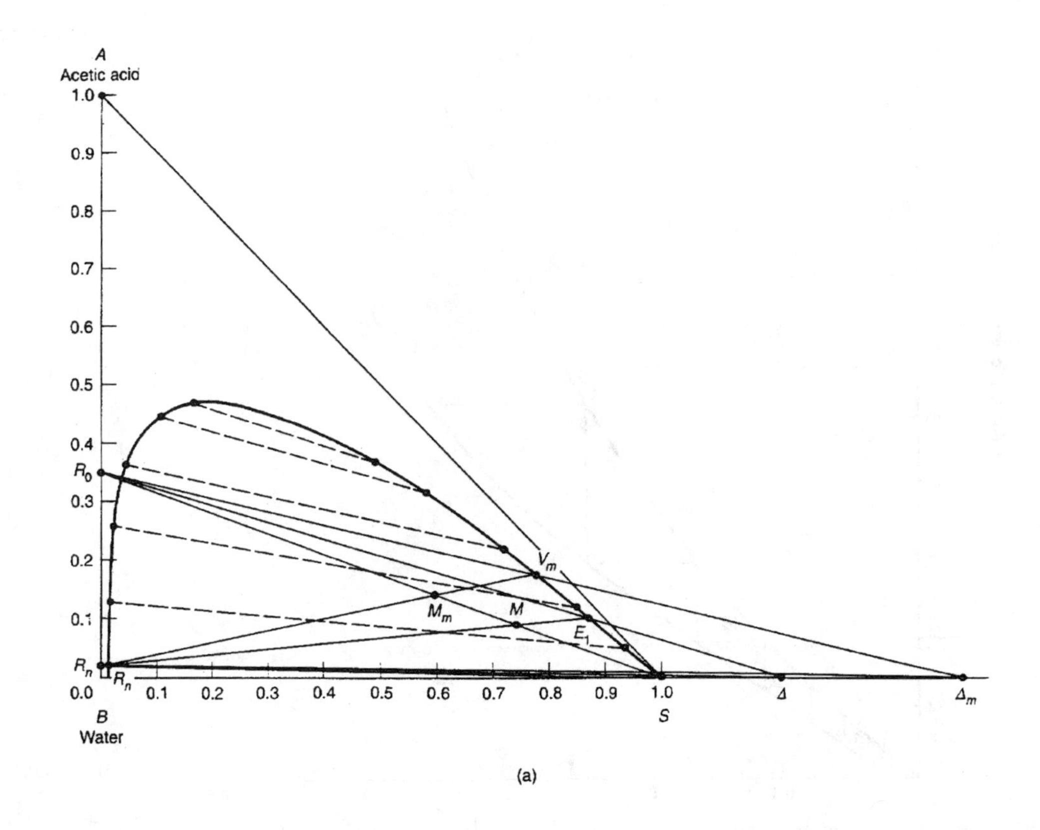

(a)

This line cuts the binodal curve in V_m and $\overline{R_nS}$ extended in Δm - difference point for minimum solvent.

Join R_n and V_m to intersect $\overline{R_0S}$ in M_m .

Minimum solvent-to-feed ratio = $\dfrac{\overline{R_0M_m}}{\overline{M_mS}} = \frac{8.1}{5.4} = 1.5185$

Actual solvent-to-feed ratio = $1.5185 \times 2 = 3.037$

Solvent to be used = $3.037 \times 5000 =$ 15185 kg/h

Solvent to be used per kg of feed solution = 3.037 kg

Calculate the composition of the addition point.

$$(x_A)_m = \frac{0.305}{4.037} = 0.087 \qquad (x_S)_m = \frac{3.037}{4.037} = 0.7523$$

Extend $\overline{R_nM}$ to cut the binodal curve in E_1

Join R_0 and E_1 and extend $\overline{R_0E}$ to intersect $\overline{R_nS}$ in the operating difference point.

Since the construction is getting crowded on the triangular diagram, it is convenient to construct equilibrium and operating lines on rectangular coordinates and step off the equilibrium stages on it. This is done in Figure **8-6b** as follows.

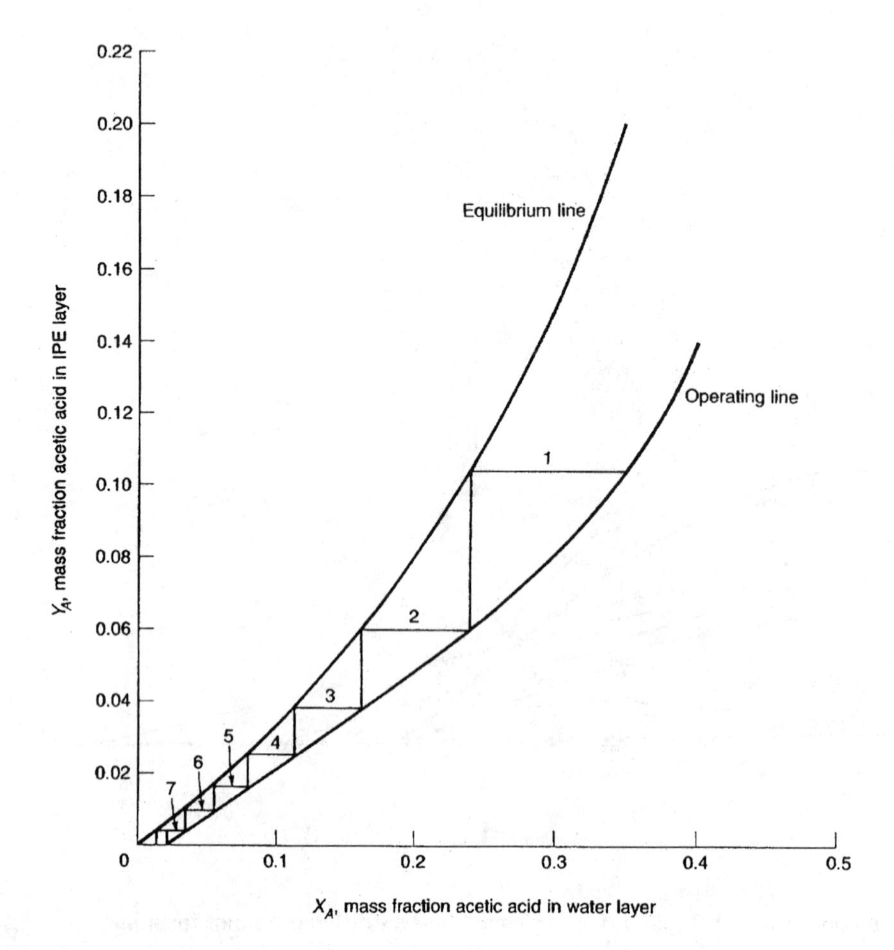

FIGURE 8-6b.

Few lines are drawn from the difference point to intersect the binodal curve and the corresponding compositions of extract and raffinate are read. These are given below.

Mass fraction , x_A	0.1	0.2	0.25	0.3	0.35	0.4
Mass fraction , x_S	0.022	0.047	0.064	0.08	0.103	0.14

These points are plotted in Figure 8-6b to give the operating line The binodal equilibrium data are directly used to plot the equilibrium line.

Stages are then stepped off. The number of stages required is found to be $\doteq 7.6$
These are theoretical number of stages. To get actual stages, a stage efficiency factor would be required.

PSYCHROMETRY AND HUMIDIFICATION

9-1: The wet-bulb temperature in a school building is 20 °C and its dry-bulb temperature is 30 °C. What is the relative humidity in the building. Assume Pr = 0.707, of air = 0.02 cP, density of air = 1.185 kg/m³ , diffusivity of water in air = 2.6 x10⁻⁵ m²/s.

9-2: An exothermic reaction system requires 60000 kg/h of cooling water at 25 °C. Water leaves the reactor jacket at 50 °C. Ambient air is at 20 °C and has 50 % relative humidity corresponding to absolute humidity of 0.007 kg water/kg dry air. The water rate must be kept at 10000 kg/h.m² minimum. The transfer coefficients are :(a) $k_y a$ = 320 kmol/h.m³ and b) $h_l a$ = 18000 W/m³. °C. A schematic of the installation is shown in Figure 9.1. Equilibrium curve- saturation humidities are plotted in Figure 9.2.
Use actual air rate = 1.5 x minimum air rate.
Prandtl number for air may be taken equal to 0.707.

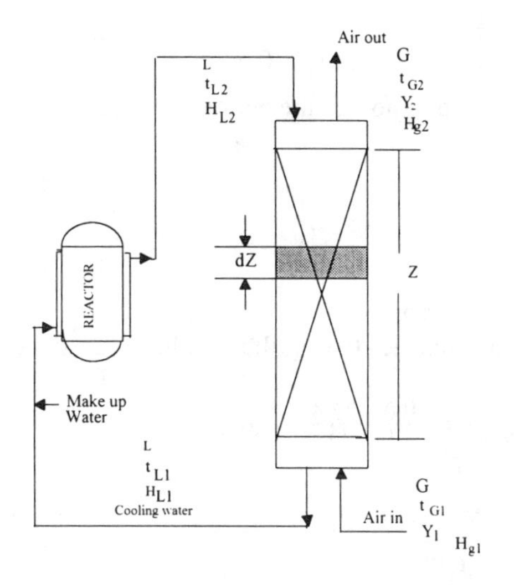

Figure 9.1: Schematics of the cooling tower serving a reactor.

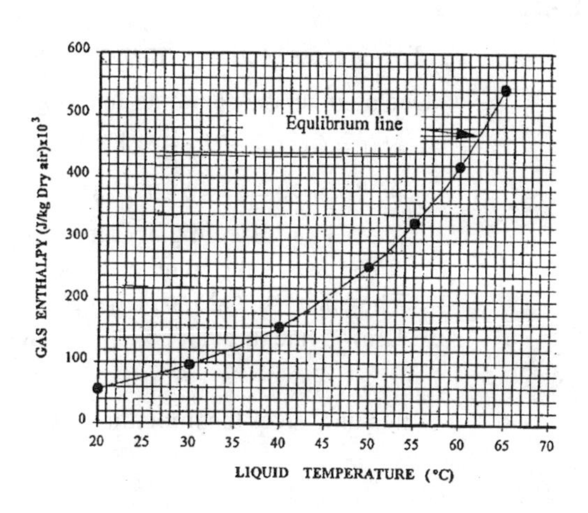

Figure 9.2: Equilibrium curve

Specific heat of air-water mixture (assumed constant) is 1010 J/kg.°C. Estimate the height of packing needed to achieve the desired cooling.

Solutions:

9-1:

Since water vapor concentration is low, use of air properties is adequate.
Average temperature =(20 + 30) /2 = 25 °C.

Schmidt number, $$\text{Sc} = \frac{\mu}{\rho D_{AB}} = 1010\frac{0.02x10^{-3}}{1.185\times 2.6\times 10^{-5}} = 0.652$$

$$\frac{h_G}{k_y} = 1010\left(\frac{Sc}{\text{Pr}}\right)^{2/3} = 1010\left(\frac{0.652}{0.707}\right)^{2/3} = 954 \text{ J/kg.°C.}$$

ΔH_v for water at 20 °C. = 2.563x10⁶ J/kg from steam properties.
vapor pressure of water at 20 °C. = 0.023 atm.

$$Y_w = \frac{0.023}{1-0.023} \times \frac{18.02}{28} = 0.0151 \quad \text{kg water vapor /kg dry air.}$$

Therefore, by substituting in wet bulb equation,

$$0{,}01515 - Y = \frac{954}{2.563x10^6}(30 - 20)$$

From which , $Y = 0.0114$ kg H_2O/kg of Dry air.
$Y' = 0.0114/0.62 = 0.0184$ mols H_2O/mol, dry air
y, mol fraction of $H_2O = 0.0184/1.0184 = 0.01807$
Then partial pressure of water = 0.01807 x 1 = 0.01807 atm
Vapor pressure of water at 30 °C = 0.041 atm
Percent humidity = (0.01807/0.041) x 100 = 44.07 %

9-2:

Tower cross section $= \frac{60000\ kg/h}{10000\ kg/h.m^2} = 6\ m^2$ Use $t_{ref} = 0$ °C

One point on the operating line is given by the condition at the bottom of tower.

$$H_{g1} = C_S\left(t_{g1} - t_{ref}\right) + Y_1'(\Delta H_v)$$

$$= 1010(20-0) + 0.007\left(2454 \times 10^3\right)$$

$$= 37.4 \times 10^3 \text{ J/kg dry air}$$

Water leaves the column at bottom at 25 °C
The point (t_{L1} = 25 °C, $H_{g1} = 37.4x10^3$) is shown as point A in Figure 9.3.
Draw a line through point A such that it just touches the equilibrium line. The slope of this line gives the minimum L/G ratio. $\frac{LC_L}{G_{Min}} = \frac{400-37.4}{70-25} = 8058$

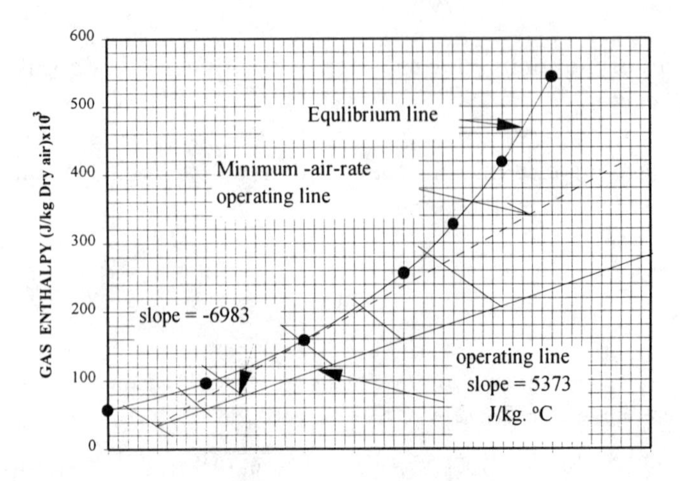

Figure 9.3 : Equilibrium curve and operating line for problem 9-2.

Therefore minimum air rate $G_M = \frac{10000\times4180}{8058} = 5187$ kg/h.m²

Actual air rate = 1.5 x 5187 – 7780 kg/h.m²

Then slope of operating line = $\frac{10000\times4180}{7780}$ = 5373 J/(kg.°C)

The operating line is drawn in Figure 9.3 with slope 5373.

To get enthalpy driving force , draw lines with slope $-\frac{h_l a}{k_x a M_B} = \frac{(1800)(3600)}{(320)(29)}$ = -6983

The values of H_i are read from the intersections of equilibrium line and a series of lines drawn with slope - 6983. and are given Table 9-1.

Table 9-1 : Enthalpy driving force calculations for problem 2

Liquid Temp °C	Gas Enthalpy, H_g (J/kg dry air)x10⁻³	Gas Enthalpy @ Interface, H_i (J/kg dry air)x10⁻³	$\frac{10^3}{H_i-H_g}$ (J/kg dry air)x10⁻¹
25	37.4	68	0.0327
30	60	86	0.0385
34	80	104	0.0417
43	120	156	0.0278
51	160	214	0 .0185
60	200	285	0.0118
68	240	362	0.0082

A plot is made showing $\frac{1}{H_i-H_g}$ vs H_g in order to evaluate the integral $\int_{H_{g1}}^{H_{g2}} \frac{dH_g}{H_i-H_g}$ in Figure

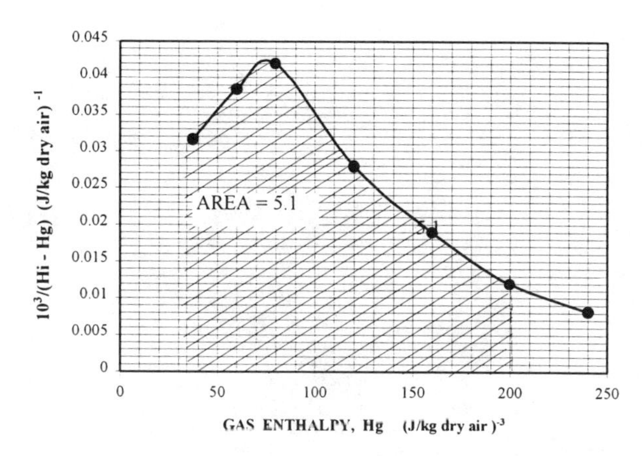

Figure 9.4. Graphical integration of enthalpy driving force for prob. 9-2.

Area under the curve is obtained by the method of counting square and is 5.1.

Therefore $\frac{k_x a M_B Z}{G} = \int_{H_{g1}}^{H_{g2}} \frac{dH_g}{H_i-H_g} = 5.1$

Now Z, the tower height can be determined as $Z = \frac{5.1(7780)}{32029} = 4.28$ m.

DRYING

10-1: A cake of a crystalline precipitate is to be dried by drawing air through it . The solid particles are 0.3 mm dia and are insoluble in water. Cake is 20 mm thick with apparent density = 85 lb/ft^3. It is to be dried from 3 to 0.1 % moisture. The air will be fed to the drier at a rate of 180 lb dry air/ft^2 bed cross section per hour and at dry-bulb temperature 90 °F and 50 % humidity. Estimate the time of drying.

10-2: The data obtained on the drying rates of a certain solid are given below.

X kg water/kg dry solid	0.35	0.2	0.18	0.16.	0.14	0.12	0.1	0.09	0.08	0.07	0.05
N kg/h.m^2	1.5	1.5	1.33	1.19	1.04	0.9	0.75	0.48	0.35	0.21	0.05

Calculate the time required to dry the cake from 30 % moisture to 5 % moisture in a batch drier if the surface available for drying is 1 m^2/35 kg of dry solid and if the drying is to be carried out under still air conditions.

10-3: There are a number of short problems given below. For each question, choose the correct answer from among the alternates listed.

10-3.1: A wet solid slab weighing 200 kg with a drying surface of 1 m^2 /75 kg is to be dried from 40 to 20 % moisture. Drying tests have shown that this drying takes place at constant rate and that the critical moisture of the solid is 20 %. The constant drying rate is $3x10^{-2}$ kg /m^2.min. The drying time (min) is closest to

a) 1500 b) 2000 c) 1050 d) 500

10-3.2:The air in a dryer is at 150 °F and has a relative humidity of 25 %. It is flowing at a velocity of 5 ft/s. The mass transfer coefficient is given by

$$k_Y = (22.4 + 0.049\,t)\,G^{0.8}$$

where G = mass velocity, lb/h.ft^2., t = dry air temperature, °F , k_Y = lb/(h)(ft^2)(unit H)
G = lb wet air/s.ft^2. The rate of evaporation (lb/h) is closest to

a) 0.2 b) 0.4 c) 0.5 d) 0.12

10-3.3 :An insoluble crystalline solid is placed in a 2 ft x 3 ft x 1 in.deep. The pan is made of 304 stainless steel, 0.32" thk . The pan is placed in an air stream at 150 °F and humidity of 0.01 lb / lb dry air.
flowing parallel to upper and lower surfaces at a mass velocity of 2300 lb/h..ft^2. The clearace between surface of the pan and an insulated pipe coil above is 4". If the pan is thoroughly insulated and radiation effect is negligible, The drying rate (lb/h) is closest to:

a) 0.976 b) 0.244 c) 3.8 d) 1.3

10-3.4: 100 pounds of a wet solid were loaded in a tray drier at a moisture content of 60 % wet basis. The overall drying rate was constant for four hours at 11 lb/h . After that the rate

decreased linearly with moisture content of the dry solid. The equilibrium moisture content of the solid is
4 %. The total moisture content in percent on wet basis 3 hours after the rate started declining is very nearly

a) 30.6 b) 40.0 c) 23.4 d) 20.0

10-3.5: An insulated hot air countercurrent drier is to be designed to dry 200 kg/h of a product containing 30 % moisture to a product containing 0.2 % moisture.(wet basis). Air will have a dry-bulb temperature of 110 °C and a wet-bulb temperature of 38°C. Air leaving the drier will be at 46 °C and 95 % saturated. The solids will enter the drier at 30 °C and will leave the drier at 60 °C. The allowable dry air velocity is 4.5 kg/min.m² of drier cross section. The dry air rate required is very nearly

a) 15.9 kg/min.m² b) 15.9 kg/min c) 953 kg/h. d) 953 kg/h.m²

Solutions to problems

10-1:

$X_1 = \frac{0.03}{1-0.03} = 0.031$ kg H_2O/kg dry solid $X_2 = \frac{0.001}{1.001} = 0.001001$ kg H_2O/kg dry solid

From the humidity chart, t_w=75 °F , Y_1= 0.0153 kg H_2O/kg dry air

Y_{As} = 0.019 kg H_2O/kg dry air

$$\text{Average } G_S = 180 + 180\left(\frac{0.0153 + 0.019}{2}\right) = 183.1 \text{ kg/h.ft}^2$$

$N_{max} = G_S (Y_{AS} - Y_1) = 183.1(0.019 - .0153) = 0.6775$ kg/h.ft²

Thickness of solid, Z_S = 20 mm = 0.06562 ft

$d_p = 9.84\text{x}10^{-4}$ ft

ρ_S = 85 lb/ft³ viscosity of air at av temp of 82.5 °F = 0.018 cP

$$\frac{d_p G_S}{\mu} = \frac{9.84\text{x}10^{-4} \times 183.1}{0.018 \times 2.42} = 4.14$$

$$N_{tG} = \frac{1.14}{d_p^{0.35}}\left(\frac{d_p G_S}{\mu}\right)^{0.215}\left(X\rho_S Z_S\right)^{0.64}$$

$$= \frac{1.14}{(9.84\times10^{-4})^{0.35}}\left(4.14\right)^{0.215}\{X \times 85 \times 0.06562)\}^{0.64}$$

$$= 52.45\,(X)^{0.64}$$

$N = N_{max}\left(1 - e^{-52.45(X)^{0.64}}\right)$ (see Mass Transfer operations, Treybal)

For estimation of θ, time of drying, the drying rate as a function of X is required.

$\int \frac{dX}{N}$ is needed to calculate θ

For quick estimate , divide the interval into two equal parts and calculate the following data

	X	N	1/N
X_1	0.031	0.675	1.4815
X_2	0.00101	0.3184	3.1407
X_{av}	0.01597	0.661	1.5129

Applying Simpson's rule, area = $\frac{0.02992}{6}[1.4815 + 4\times 1.5129 + 3.141] = 0.05322$

$$\frac{L_S}{A} = \left(\rho_S Z_S\right) = 85(0.06562) = 5.5777$$

Then $\theta = \frac{L_S}{A}\int \frac{dX}{N}$ = $5.777(0.05322) = 0.297$ h = 17.8 min

10-2:

Plot the graph of N vs X . This is done in figure 10-1.
Next calculate time for the duration when rate of drying is constant.

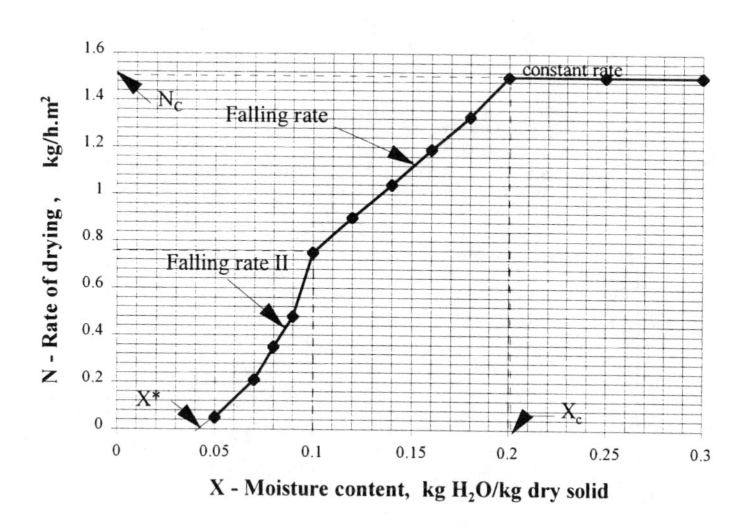

Figure 10-1: Rate of drying vs X

Time for constant drying rate period

From the graph, X_C = 0.2 , N_C = 1.5 kg/m².h

X_1 = 0.3/0.7 = 0.4286 kg H_2O/kg dry solid

X_2 = (.05/0.95) = .05263 kg H_2O/kg dry solid

$X^* = 0.043$ kg H_2O/kg dry solid
N_C = drying rate kg/h.m²

(1) θ_c

$$\theta_C = \frac{L_S}{AN_C}(X_1 - X_C) = \frac{35}{1.5}(0.4286 - 0.2) = 5.334 \text{ h}$$

(2) θ_{f1} **Time for falling rate period I**

In this case $N_1 = N_c = 1.5$ kg/h.m² $N_2 = 0.75$ kg/h.m²

$$\theta_{f1} = \frac{L_S}{A}\left(\frac{X_C - X_2}{N_1 - N_2}\right)\ln\left(\frac{N_1}{N_2}\right) = 35\left(\frac{0.2-0.1}{1.5-0.75}\right)\ln\frac{1.5}{0.75} = 3.24 \text{ h}$$

(3) **Falling rate period II**

In this period rate curve is not a straight line and time is to be obtained by graphical integration . For this, values of 1/N are computed and plotted against X to get the integral $\int \frac{dx}{N}$. This plot is shown in Figure 10-2 below.

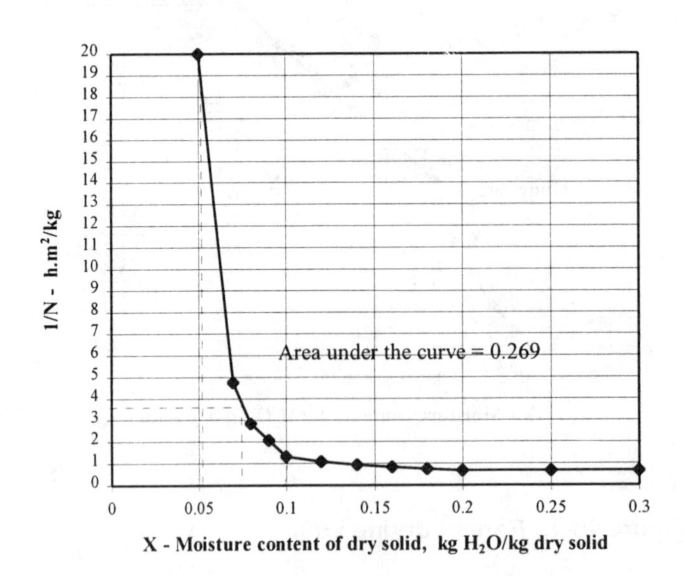

Figure 10-2: Graphical integration for second falling rate perioe in Prob. 10-2.

The area under the curve is found by Simpson's rule to be 0.2688.

$$\theta_{f2} = \frac{35}{1}\int\frac{dX}{N} = 35 \times 0.2688 = 9.41 \text{ h}$$

Total time of drying = 5.33 + 3.24 + 9.41 = 17.98 h

10-3:

10-3.1:

For the period when the drying rate is constant , the time of drying is given by

$$\theta c = \frac{L_S(X_1 - X_2)}{A N_c}$$

$L_S/A = 75$ kg/m², $N_c = 3\text{x}10^{-2}$ kg /(m².min)

$$X_1 = \frac{0.4}{1-0.4} = 0.67 \qquad X_2 = \frac{0.2}{1-0.2} = 0.25$$

$$\theta_c = 75 \times \frac{0.67-0.25}{3\times10^{-2}} = 1050 \text{ min}$$

Answer is c)

10-3.2:

From humidity chart,

Air at 150 °F and 25 % relative humidity, H = 0.0415 lb H_2O /lb Dry air.

Also wet-bulb temperature = 106 °F. and H_S = 0.0525. lb H_2O /lb Dry air.

Humid volume, $v_H = 359\left(\frac{1}{29} + \frac{0.0415}{18.02}\right)\left(\frac{610}{492}\right) = 16.37$ ft³ /lb

G = 5 x1/16.37 = 0.305 lb/s.ft²

$k_Y = [22.4 + 0.049 \times 150]\,(0.305)^{0.8} = 11.1$ lb/h.sq.ft.(unit H)

Then the rate of evaporation = 11.1(0.0525 - 0.0415) = 0.122 lb/h..ft²

Answer is d)

10-3.3 : Since conduction and radiation effects are absent, the solid drying surface attains wet-bulb temperature. For air-water vapor mixture $t_w = t_s$

For air at 150 °F & 0.01 humidity, wet- bulb temperature of air = 84 °F (from humidity chart).

Y_S = 0.025 lb/lb of dry air, λ_S = 1046.8 btu/lb

Calculation of h_c :

equivalent diameter = $\frac{4\times4/12)\times2}{2(4/12+2)} = 0.57$ ft

$$h_c = (0.0135)\frac{2300^{0.71}}{d_e^{0.29}} = 0.0135\left(\frac{2300^{0.71}}{0.57^{0.29}}\right) = 3.8722 \text{ Btu/(h.ft}^2.^\circ\text{F)}$$

$$N_c = \frac{3.8722(150-84)}{1046.8} = 0.24 \text{ lb/ft}^2\text{.h}$$

Evaporation rate = 0.244(2 x 2) = 0.976 lb/h

Answer is a)

10-3.4:

The constant rate period lasts for 4 hours

Therefore, water evaporated in 4 hours = 4x11 = 44 lb

Water remaining with the solid after 4 hours = 100x0.6 - 44 = 16 lb

Critical moisture content = X_C = 16 /40 = 0.4 lb H_2O/ lb dry solid.
Moisture content at the start = X_1 = 60/40 = 1.5 lb H_2O/ lb dry solid
If A is the tray area used for loading the solid , we can write

$$\theta_c = \frac{100}{A} = \frac{(X_1 - X_c)}{N_c}$$

$$\text{Therefore } AN_C = \frac{100(X_1 - X_c)}{\theta_c} = \frac{100(1.5 - 0.4)}{4} = 27.5$$

Assume during falling rate period, the drying rate is proportional to moisture content X.
Equilibrium moisture content = $\frac{0.4}{0.96}$ = 0.0417 lb H_2O/ lb dry solid
Let X_2 be the moisture 3 hours after the drying rate begins to fall. θ_{f1} = 3 hours
Then

$$\theta_{f1} = \frac{L_S}{A}\frac{(X_c - X*)}{N_c}\ln\frac{X_c - X*}{X_2 - X*} = (27.5)(0.4 - 0.0417)\ln\frac{(0.4 - 0.0417)}{(X_2 - 0.0417)} = 3$$

Solving for X_2 gives X_2 = 0.306 lb H_2O/ lb dry solid
Therefore % water in solid = (0.306 / 1.306)x 100 = 23.4 % **Answer is c)**

10-3.5:

$$X_1 = \frac{0.3}{1 - 0.3} = 0.4286 \text{ lb } H_2O/\text{ lb dry solid} \quad X_2 = \frac{0.002}{1 - 0.002} = 0.002$$

L_S = 200(1 - 0.4286) = 114.3 kg/h dry solid flow.
Water evaporation per hour = (0.4286 - 0.002)(114.3) = 48.76 lb/h
Absolute humidity of entering air = 0.125 kg water/ kg dry air.
Absolute humidity of air leaving the drier = Y_1 , to be determined.
Air is to leave the drier at 46 °C 95 % saturated
Saturation humidity at 46 °C = 0.067 kg H_2O/kg dry air
Therefore Y_1 = 0.95(0.067) = 0.06365 kg H_2O/kg dry air
Change of humidity per lb of dry air = 0.06365 - 0.0125 = 0.05115 kg H_2O/kg dry air
dry air rate = 48.76/0.05115 = 953.3 kg/h.= 15.9 kg/min.

Answer is b)

FILTRATION

11-1: A homogeneous slurry is filtered through a batch filter at constant pressure difference of 50 psi. The cake formed is incompressible 3/4" cake is formed in 50 min with a filtrate volume of 1500 gal. 3 minutes are required to drain the filter, 2 minutes to fill the filter with water washing is carried out in the same manner as filtration using 300 gallons of water. Opening, dumping and closing the filter takes 5 minutes. Assume that filtrate has the same properties as wash water and neglect the resistance's of the filter cloth and the filter channels.

(a) How many gallons of filtrate is produced on the average per 24 hours?
(b) How much filtrate could be produced if a cake of 5/8" thickness is formed using the same ratio of wash water to filtrate with other conditions remaining the same.
(c) Determine the optimum thickness and maximum daily capacity of filtrate production for the conditions in above.

11-2: A leaf-type filter having filtration area of 2 ft^2 was used to obtain the following data on filtration of a compressible slurry at constant rate of filtration of 1 gpm.

Pressure drop psi	10	20	30	40	50	60	75	90
filtrate (gal)	3.7	7.0	10	12.8	15.6	18.4	22.3	26.4

(a) If this filter is operated on the slurry at constant rate of filtration until the pressure drop becomes 65 psi, and then a constant-pressure operation until 100 gal filtrate is obtained, then washed at 65 psi with 20 gal of wash liquor and dumped in 30 min, how much time will be required to produce 1000 gal of filtrate? Assume that the resistance of the filter itself is negligible.

(b) A rotary drum filter 6 feet in diameter and 6 feet in width rotating at 6 min per revolution is to be used to filter the same slurry. The drum is submerged in the slurry 30% and the pressure inside the drum is 4.7 psia. Assuming the resistance of the filter is negligible, how much time will be required to produce 1000 gal of filtrate?

11-3: The filtration data obtained on a certain slurry at a constant pressure drop of 49.1 psi when plotted as $\frac{\Delta\theta}{\Delta V}$ *vs* V on rectangular coordinates gave a straight line with a slope of 4874 s/ft^6 and an intercept of 177.86 s/ft^3. The following other data are also available:

solids concentration in the slurry = 1.47 lb/ft^3,
density of solids = 164.1 lb/ft^3,
density of dry cake = 73.5 lb/ft^3,
mass ratio of wet cake to dry cake = 1.6
area of filter = 0.474 ft^2

Choose the correct answer for each of the following questions.

11-3.1 The mass (lb/lb) of solids per unit mass of filtrate is closest to

a) 0.0236 b) 0.0239 c) 1.47 d) 1.49

11-3.2 The specific resistance of the cake α $\frac{s^2 lb_f}{l120b}$ is closest to

a) 8.72×10^9 b) 2.8×10^1 c) 8.7×10^{10} d) 2.8×10^{10}

11-3.3 The following data were also obtained for specific resistance as a function of pressure drop:

pressure drop psi	6.1	16.2	28.2	36.3
specific resistance $\alpha\times10^{-9}$ $\left(\frac{s^2\cdot lb_f}{lb_m^2}\right)$	5.17	6.71	7.8	8.18

The compressibility of the cake is closest to

a) 0.4 b) 0.5 c) 0.26 d) 0.7

11-3.4 Equivalent volume (ft³) of filtrate corresponding to the filter medium resistance is closest to

a) 0.1 b) 0.0365 c) 0.5 d) 0.01825

11-3.5 The filter medium resistance $\left(\frac{h^2\cdot lb_f}{lb_m\cdot ft^2}\right)$ is closest to

a) 1060 b) 1200 c) 7363 d) 550

11-3.6 The porosity of the cake is closest to

a) 0.3 b) 0.60 c) 0.55 d) 0.47

11-3.7 The permeability of the cake $\left(\frac{lb_m\cdot ft^3}{lb_f s^2}\right)$ is closest to

a) 1.6×10^{-5} b) $1.6\times10^{-}$ c) 5.3×10^{-3} d) 2.7×10^{-4}

11-3.8 The time (s) required to obtain 0.2 ft³ at a constant pressure drop of 49.1 psi is closest to

a) 230 b) 133 c) 115 d) 300

11-3.9 At the same conditions as in 8 above, when the filtrate collected is 0.15 ft³, the rate of filtration (ft³ /s) is closest to

a) 0.011 b) 0.1 c) 0.005 d) 0.0011

11-3.10 Under the same conditions as in 8 above, assuming the porosity of the cake is constant, the cake thickness (cm) when the filtrate collected is 0.2 ft³, is closest to

a) 0.26 b) 0.1 c) 2.6 d) 1.3

Solutions to problems

11-1 :

(a) For constant ΔP, (and incompressible cake) $\theta = \frac{C_V V^2}{A^2 \Delta P}$

$$\frac{A^2}{C_V} = \frac{V^2}{\theta \Delta P} = \frac{1500^2}{50 \times 50} = 900$$

Final filtration rate when total filtrate volume is 1500 gal = $\frac{dV}{d\theta} = \frac{A^2}{C_V}\left(\frac{\Delta P}{2V}\right)$

$$= 900 \times \frac{50}{2 \times 1500} = 15 \; gallons/\text{min}$$

Thus rate of washing is 15 gal/min. Then time of washing = 300/15 = 20 min.
Then cycle time

	Min
Filtration	50
Drain	3
Fill	2
Wash	20
Drain	3
Dump	5
Fill	2
Total	85 minutes.

Average filtrate production per 24 hours = $\frac{24 \times 60}{85} \times 1500 = 25412 \; gal$

(b) The filtrate is directly proportional to cake thickness .

Therefore, filtrate = $\frac{5/8}{3/4} \times 1500 = 1250$ gal. in a batch.

Wash water = $\frac{1250}{1500} \times 300 = 250 \quad gal/batch.$

Filtration rate $\frac{dV}{d\theta} = \frac{A^2}{C_V}\left(\frac{\Delta P}{2V}\right) = 900 \frac{50}{2 \times 1250} = 18 \quad ga$ when 1250 gal of filtrate are collected.

$$\text{Filtration time} = \frac{C_V}{A^2}\frac{V^2}{\Delta P} = \frac{1250^2}{900 \times 50} = 34.72 \text{min}$$

Wash time = 250/18 = 13.9 min

cycle time

	Min
Filtration	34.72
Drain	3.0
Fill	2.0
Wash	18.9
Drain	3.0
Dump	5.0
Fill	2.0
Total	68.62

Daily filtrate production = (24x60/68.62) x 1250 = 26231 gal

(c) Let V be the volume of filtrate per batch corresponding to maximum daily filtrate production of V_D

This also corresponds to the optimum thickness

Time of filtration = $\theta_f = \frac{C_V}{A^2}\frac{V^2}{\Delta P} = V^2/(50 \times 900)$

Final rate of filtration = $\frac{dV}{d\theta} = \frac{A^2}{C_V}\frac{\Delta P}{2V} = 900 \times 50/(2 \times V)$

Wash water = (300/1500)x V = 0.2V

Time of washing = $\frac{0.2V}{\frac{900\times50}{2V}} = \frac{0.4V^2}{900\times50}$

Other time as in a = 3 +2 + 5 +3 +2 =15 min.

Therefore an expression for total average daily filtrate production can be written as follows

$$V_D = \frac{24 \times 60V}{\frac{V^2}{900\times50} + \frac{0.4V^2}{900\times50} + 15} = \frac{24\times60V}{\frac{1.4V^2}{900\times50} + 15}$$

For maximum V_D, $\frac{dV_D}{dV} = 0$

Therefore differentiating the expression for V_D and equating it to 0

$$\frac{dV_D}{dV} = \frac{\left(\frac{1.4V^2}{900\times50} + 15\right)(24\times60) - 24\times60V\left[\frac{2.8V}{900\times50}\right]}{\left(\frac{1.4V^2}{900\times50} + 15\right)^2} = 0$$

For this to be true, the numerator must be equal 0. Therefore

By equating the numerator to 0 , one gets $\frac{1.4V^2}{900\times50} - 15 = 0$

or $V^2 = 15 \times 900 \times 50/1.4 = 482142.86$ from which V = 694.37 gal per batch.

Optimum thickness = (694.37/1500)(3/4) = 0.347"
Time of filtration = (694.37)²/(900x50) = 10.71 min.
Time of washing = 0.4x(694.37)² /(900x50) = 4.28 min
Total cycle time = 10.71 + 4.28 + 15 = 29.99 ≑ 30 min
Then maximum daily filtrate = (24x60/30)(694.37) = 33329.76 ≑33330 gal per day.

11-2:

At constant rate of filtration, and no filter resistance,

$$\Delta P^{1-s} = \frac{2C'_V V}{A^2}\frac{dV}{d\theta}$$

Since $\frac{dV}{d\theta} = 1$ gpm, the above reduces to $\Delta P^{1-s} = \frac{2C'_V V}{A^2}$

Taking logarithms of both sides, $(1-s)\ln \Delta P = \ln V + \ln \frac{2C'_V}{A^2}$

or $$\ln \Delta P = \frac{1}{1-s} \ln V + \frac{1}{1-s} \ln \frac{2C'_V}{A^2}$$

Thus a plot of $\ln \Delta$ vs $\ln$ will be a straight line with slope 1/(1 - s) and intercept

of $\frac{1}{1-s} \ln \frac{2C'_V}{A^2}$ This plot is shown in figure 11-1.

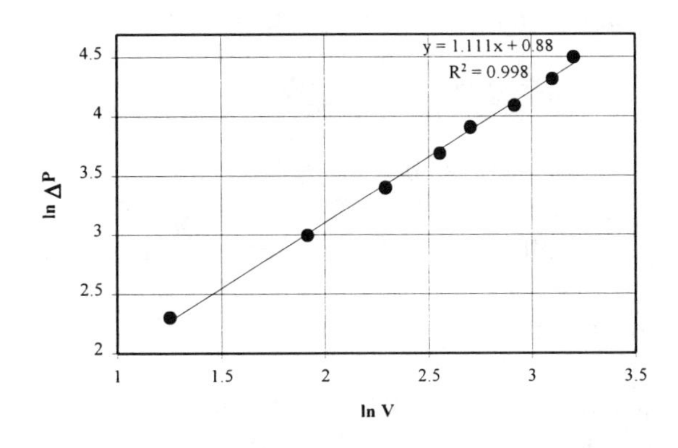

Figure 11-1 : Plot of ln ΔP vs ln V for problem 11-2

The slope of this line is 1.111 and intercept is 0.88

Compressibility of the cake

$\frac{1}{1-s} = 1.11$ s = 1/1.111 = 0.9 Therefore s = 1 - 0.9 = 0.1

Calculation of C'_V

$\frac{1}{1-s} \ln \frac{2C'_V}{A^2} = 0.88$ then $\frac{2C'_V}{A^2} = e^{0.88(1-s)} = e^{0.88 \times 0.9} = 2.20$

Since A = 2 ft², $C'_V = 4.41$

Volume of filtrate and time for constant rate filtration upto 65 psi pressure drop:

$$\Delta P^{(1-s)} = \Delta P^{(1-0.1)} = \frac{2C'_V}{A^2} V \times \frac{dV}{d\theta}$$

Substitution of appropriate known values in the above gives

$$(65)^{0.9} = 2.208 \times V \times 1$$

From which V = 19.4 gal. Since rate is constant at 1 gpm,
time of constant rate filtration = 19.4 min.

Time of constant-pressure filtration:

$$\theta = \frac{C'_V}{A^2}\frac{V^2}{(\Delta P)^{(1-s)}} = \left[\frac{2.208}{2}\frac{V^2}{(65)^{0.9}}\right]_{19.4}^{100} = \frac{2.208}{2\times 65^{0.9}}\left[100^2 - 19.4^2\right] = 248.1 \text{ min}$$

Time of washing:

Filtration rate at the end of filtration is given by

$$\frac{dV}{d\theta} = \frac{A^2}{2C'_V}\frac{\Delta P^{0.9}}{V} = \frac{1}{2.208}\frac{65^{0.9}}{100} = 0.194 \text{ gpm}$$

Therefore time of washing = 20/0.194 = 103.1 min
Total filtration cycle = 30 + 19.4 + 248.1 + 103.1 = 400.6 min.
This is required to collect 100 gal. Therefore, time to collect 1000 gal = 4006 min

Rotary filter filtration:

Time of cake deposition = 6(0.3) = 1.8 min
Area of cake deposition = $0.3(\pi DL) = 0.3(\pi \times 6 \times 6) = 33.93$ ft²
$\Delta P = 14.7 - 4.7 = 10$ psi

For constant pressure filtration, $\theta = \frac{C'_V}{A^2}\frac{V^2}{(\Delta P)^{(1-s)}}$

Therefore, $1.8 = \frac{4.416}{33.93^2} \times \frac{V^2}{10^{0.9}}$

and $V^2 = \frac{10^{0.9} \times 33.93^2 \times 1.8}{4.416} = 3727.45$

Then volume of filtrate collected = V = $\sqrt{3727.45} = 61.05$ gal
Total cycle time is 6 min
Therefore, time for collecting 1000 gal of filtrate = (1000/61.05)(6) = 98.28 min

11-3 :

11-3.1 mass of solids/unit of filtrate = $\frac{1.47 \; lb/ft^3}{1-[1.6-1]\frac{1.47}{62.4}} = 1.491$ lb/ft³

Therefore, mass ratio of solids deposited to unit mass of filtrate = 1.4911/62.4 = 0.0239
answer is b)

11-3.2 slope of line = $4874 = \frac{2C_V}{A^2 \Delta P}$ and

$$a = \frac{2C_V}{r\rho\mu} = \frac{4873A^2\Delta P}{(0.0239)(62.4)(0.886\times0.000672)} = 8.72\times10 \quad \left(\frac{s^2\cdot lb_f}{lb_m^2}\right)$$

answer is a)

11-3.3 $a = a_0(\Delta P)^s$ Therefore, a plot of a vs Δ on log-log paper or ln a vs ln Δ on rectangular coordinates should yield a straight line with slope = s. The values of ln a are plotted against the values of ln(ΔP) in the following figure.

ln α (Specific resistance) vs ln ΔP (Pressure drop)

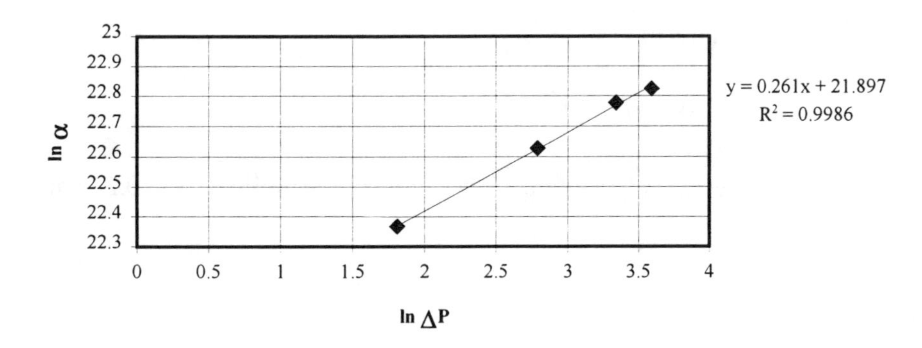

The plot is straight line with slope = 0.261. Thus compressibility of the cake is 0.261

answer is c)

11- 3.4 Intercept of st. line on plot of $\frac{\Delta\theta}{\Delta V}$ *vs* V = 177.86 (given in the problem)

$$= \frac{2C_V}{A^2\Delta P}\times V_e$$

$$V_e = \frac{intercept}{slope}$$

Then $V_e = \frac{177.86\times A^2\Delta P}{2C_V} = \frac{177.86A^2\Delta P}{(slope)(A^2\Delta P)} = \frac{177.86}{4874} = 0.036$ ft³

Answer is b)

11-3.5 Filter medium resistance = $\frac{V_e\rho a}{A} = \frac{0.0365\times62.4\times8.72\times10^9/3600}{0.474}$

$$= 7363.6\left(\frac{h^2\cdot lb_f}{ft^2\cdot lb_m}\right)$$

Answer is c)

11-3.6 Porosity of the cake

$(1-X)\rho_s = 73.5$ from which X = 0.552

Answer is b)

11-3.7 Permeability of the cake

$$K = \frac{1}{\alpha \rho_s (1-X)} = \frac{1 \times 3600^2}{8.72\times10^9 \times 164.1 \times (1-0.552)} = 1.64 \times 10^{-} \quad \frac{lb_m ft^3}{lb_f s^2}$$

Answer is a)

11-3.8 Time required to obtain 0.2 ft³ of filtrate

$$\theta = \frac{C_V(V^2+VV_e)}{A^2 \Delta P} = \frac{\frac{1}{2}(4874)A^2\Delta P)}{A^2 \Delta P}(0.2^2 + 2 \times 0.2 \times 0.0365) = 133.06$$

Answer is b)

11-3.9 The rate of filtration

$$\frac{dV}{d\theta} = \frac{A^2 \Delta P}{2C_V(V+V_e)} = \frac{A^2 \Delta P}{4874A^2 \Delta P(0.15+0.0365)} = 0.001 \quad ft^3/s$$

Answer is d)

11-3.10 The thickness of cake, L

$$(1-X)LA\rho_s = V\rho$$

$$L = \frac{V\rho r}{(1-X)A\rho_s} = \frac{0.2 \times 62.4 \times 0.0239}{(1-0.522)(0.474)(164.1)} = 0.0085 \quad ft = 0.26 \text{ cm}$$

Answer is a)

THERMODYNAMICS

12-1: A steam turbine operates under the following conditions. Adiabatic operation. Steam to turbine at 900 °F and 700 psia. Steam exhausts at 60 psia. Calculate the amount of work that can be obtained from this turbine if its efficiency is 85 % and the steam rate is 30000 lb/hr. What is the entropy increase per lb of steam and what is the temperature of the steam exiting the turbine? When power demand is low, power output of the turbine is reduced by throttling steam to turbine to a lower pressure of 450 psia. Compute the steam temperature to the turbine, temperature of steam exiting the turbine and the reduced output of the turbine in this case.

12-2: Ethane is compressed from 100 psia and 80°F to 1000 psia in a two stage compressor with interstage cooling to 120 °F. Calculate the work for this compression if the overall efficiency of compression is 80 % and the flow of ethane is 100 lb/min. What is the cooler duty and cooling water required in gpm if cooling water is available at 86°F and cooling water return is at 104 °F? The following property data are available for ethane:

Properties of ethane

100 psia				300 psia				1000 psia			
Temp.		h	s	Temp.		h	s	Temp.		h	s
°F	ft³/lb	Btu/lb	Btu/(lb°F)	°F	ft³/lb	Btu/lb	Btu/(lb°F)	°F	ft³/lb	Btu/lb	Btu/(lb°F)
80	1.823	453.5	1.6946	120	0.5993	459.7	1.6396	200	0.1953	470.7	1.6102
90	1.863	458.0	1.7030	180	0.6903	491.0	1.6916	220	0.2100	484.7	1.6211
100	1.902	462.5	1.7111	200	0.7193	501.6	1.7080	240	0.2238	498.1	1.6405

Solutions to problems:

12-1:

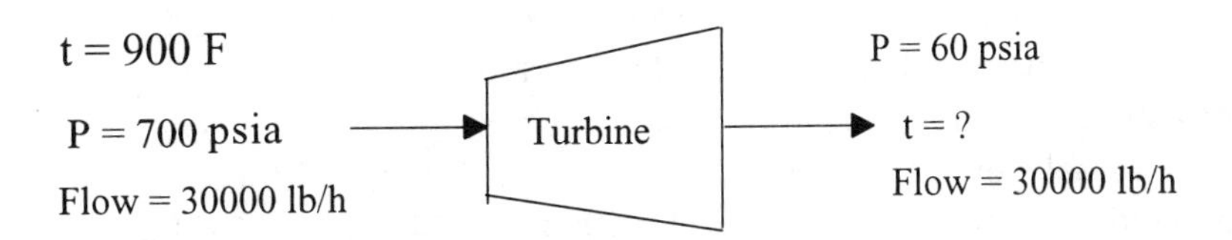

Mass balance on the control volume bounded by the turbine as system reduces to $M_{II} = M_o$

The energy balance can be written as follows

$$\overset{1}{(U + PE + KE)_E} - \overset{1}{(U + PE + KE)_B} = \Sigma(H + \overset{2}{PE} + KE)_I - \Sigma(H + \overset{2}{PE} + KE)_o + \overset{3}{Q} - W + \overset{4}{E_p}$$

1 Steady state
2. Assume negligible KE and PE effects
3. Q = 0 because of adiabatic operation.
4. No atomic level transmutations.

The energy balance reduces to $0 = H_I - H_o - W$ or $W = M(\underline{H}_I - \underline{H}_o)$

From steam tables, inlet conditions are as follows

P = 700 psia, t = 900 °F, $\underline{H}_I$ = 1458.2 Btu/lb, $\underline{S}_I$ = 1.6567 Btu/lb·°R
For maximum work, the process is internally reversible and therefore S_P = 0
Further the entropy balance reduces to $\underline{S}_I = \underline{S}_0$
Outlet conditions:
Again using steam tables and interpolating at 60 psia and $\underline{S}_0$ = 1.6567 Btu/lb°R

$$t_o = (0.0103/0.037)(50) + 300 = 313.92 \sim 314\ °F$$

$$\underline{H}_o = (0.0103/0.037)(26.7) + 1181.8 = 1191.2\ Btu/lb$$

Therefore maximum work = $\underline{H}_I - \underline{H}_o$ = 1458.2 - 1191.2 = 267 Btu/lb
Actual work = 0.85(267) = 227 Btu/lb. Then actual $\underline{H}_o$ = 1458.2 - 227 = 1231.2 Btu/lb
Again by interpolation, t_a = (22.7/25.5) (50) + 350 = 394.5 °F

$\underline{H}_o$ (actual) = 1231.2 Btu/lb (calculated above)
$\underline{S}_0$ (actual) = 922.7/25.5)(0.0305) + 1.6834 = 1.7106 Btu/lb°R

Total actual work output of the turbine = 227(30000/60) = 113500 Btu/min = **2675 hp**

Case 2:
In this case steam pressure is reduced to 450 psia by throttling. The throttling is isenthalpic process. Therefore, the properties of steam at 450 psia can be found from steam tables by interpolation as follows

t_I = (44.8/53.6)(100) + 800 = 883.6 °F
$\underline{S}_i$ = (44.8/53.6)(0.0409) + 1.6694 = 1.7036 Btu/lb°R
At 60 psia and $\underline{S}_i$ = 1.7036 Btu/lb°R, t = (0.0202/0.0305)(50) + 350 = 383.1 °F
and $\underline{H}_o$ = (0.0202/0.0305) (25.5(+ 1208.5 = 1225.4 Btu/lb
Max. work = 1458.2 - 1225.4 = 232.8 Btu/lb
Assuming the same efficiency as in first case,
actual work output of turbine = 0.85(232.8) = 197.9 Btu/lb
Total reduced actual work = (197.9 x 30000/60)/ 42.427 = **2278.5 hp.**
Actual $\underline{H}_0$ = 1458.2 - 197.3 = **1260.3** Btu/lb and t_o = (2.6/24.1)(50) + 450 = 454.4 °F

12-2:

First calculate maximum work in both stages.

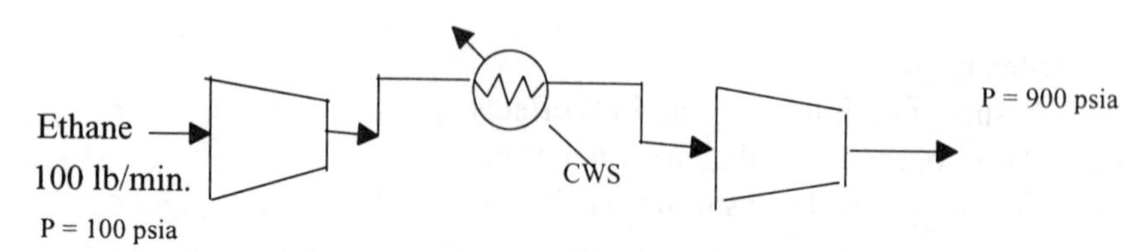

Two stage compression with intermediate cooling

For maximum work, the process is reversible, and $\Delta S = 0$
From tables, properties of ethane at inlet conditions are

t = 80 ^{0}F P = 100 psia H = 435.5 Btu /lb S = 1.6946 Btu/lb ^{0}F

Energy balance reduçes to W = M ($\underline{H}_i$ - $\underline{H}_o$)

The compression ratio = $\sqrt{\frac{P_3}{P_1}} = \sqrt{\frac{900}{100}} = 3$ Therefore , P_2 = 100 x 3 = 300 psia

at outlet of first stage, the conditions are

P = 300 psia S = 1.6946 Btu/lb ^{0}R

$$\underline{H} = 491 + (501.6 - 491.0) \times \frac{(1.6946 - 1.6916)}{(1.7080 - 1.6916)} = 492.2 \text{ Btu / lb}$$

$$t = 180 + (200 - 180) \times \frac{(1.6946 - 1.6916)}{(1.7080 - 1.6916)} = 183.7\ ^0\text{F}$$

The max. work of first stage = 435.5 - 492.2 = - 56.7 Btu/lb

(- sign indicates work done on the system)

After first stage ethane is cooled to 120 ^{0}F . The properties of ethane at this temperature and 300 psia are:

$\underline{H}_2$ = 459.7 Btu/lb $\underline{S}_2$ = 1.6396 Btu /lb^{0}R

At outlet of second stage, $\underline{S}_3$ = $\underline{S}_2$ = 1.6396 Btu /lb^{0}R

$$t_3 = 220 + 20 \times \frac{1.6396 - 1.6210}{1.6405 - 1.6210} = 239.1\ ^0\text{F}$$

$$\underline{H}_3 = 484.7 = (498.1 - 484.7)\left(\frac{1.6396 - 1.621}{1.6405 - 1.621}\right) = 497.5 \text{ Btu/lb}$$

Max. work of second stage = 459.7 - 497.5 = -37.8 Btu/lb

Total max. work done = - (56.7 + 37.8) = -94.5 btu/lb

Actual work = - 94.5 x 100/0.8 = - 11813 Btu/lb

hp = 11813/42.427 = 278.4

Cooler duty:

Q = 100 x 60 x (492.2 - 459.7) = 195000 btu/hr

Cooling water required = $\frac{195000}{(104 - 86)(1)}$ = 10833 lb/h = 21.7 gpm

CHEMICAL KINETICS

13-1: Given the data in the table below, calculate the equilibrium constant and equilibrium conversion for the reaction : $C_2H_6(g) \rightarrow C_2H_4(g) + H_2(g)$ if the reaction takes place at 1000 K and 1 atm. pressure.

Table - Kinetics 1: Data on $\Delta G^0_{298}, \Delta H^0_{298},$ *and* C_P Values

Component	ΔG^0_{298}	ΔH^0_{298}	C_P
	kcal/g-mol	kcal.g-mol	kcal / g-mol.K
C_2H_6	- 7.86	- 20.23	$C_P = 0.0023 + 2\times 10^{-5}T$
C_2H_4	16.28	12.5	$C_P = 0.0028 + 3\times 10^{-5}T$
H_2	0	0	$C_P = 0.0069 + 0.4\times 10^{-5}T$

13-2: 250 g-mols of P are to be produced hourly from a feed consisting of a solution of A (Initial concentration C_{AO} = 0.2 mol/liter) in a mixed-flow reactor. The reaction is A →P. The reaction rate is given by

$$-r_A = 0.25C_A \ h^{-1}$$

Other data are as follows

Cost of reactant = \$0.9/g-mol of A . Cost of reactor including installation = \$ 0.02 /h liter What reactor size, feed rate and conversion should be used for optimum operation? What is the unit cost of P from this operation if unconverted A is discarded.

13-3: The gas phase decomposition A → B + 2C was conducted in a constant volume reactor. Data collected in various runs is given in Table-Kinetics 2. Runs 1 to 5 were conducted at 100 ^{0} and run 6 was carried out at 110 ^{0} . (a) Determine the order of the reaction and the reaction rate constant (b) What are the values of activation energy and the frequency factor for this reaction.

Table Kinetics 2: Half-Life $t_{1/2}$ as a Function of Initial concentration C_{AO}

Run No	**1**	**2**	**3**	**4**	**5**	**6**
C_{AO} **g-mol/L**	**0.025**	**0.0133**	**0.01**	**0.05**	**0.075**	**0.025**
Half-Life $t_{1/2}$ **min**	**4.1**	7.7	**9.8**	**1.96**	**1.3**	**2.0**

13-4: The hydrolysis of acetic anhydride follows the reaction

$$(CH_3CO)_2O + H_2O \rightarrow 2CH_3COOH$$

The reaction is pseudo-first-order. It is carried out in a dilute solution of anhydride between 10-40 ^{0}C . The reaction velocity constant at 25 ^{0}C is given by **k =0.158 min^{-1}**. Activation energy for this reaction is 10600 cal/g-mol. Estimate the size of a plug-flow reactor if it is to produce 300 kg/h of acetic acid from a feed containing 0.1 M anhydride. The temperature of reaction is 40 ^{0}C.

13-5: A reaction **A →R** is carried out in batch reactor. The rates of reaction as a function of the concentration are given in the table below.

C_A (g-mol/L)	0.1	0.2	0.3	0.4	0.5	0.6	0.7	1.3	2.0
$-r_A$ g-mol/L.min	0.1	0.3	0.5	0.6	0.5	0.25	0.1	0.045	0.04

(a) How long a batch must be reacted to reach a concentration of 0.3 mol/L from an initial concentration of 1.5 mol/L

(b) Using least square estimate, obtain an expression for the rate of reaction from the above data.

Solutions to problems :

13-1:

Calculation of equilibrium constant:

First determine ΔC_P of products and reactants.

$$\Delta C_P = 0.0028 + 3.0\times10^{-5}T \; + 0.0069 + 0.4\times10^{-5}T - 0.0023 - 2.0\times10^{-5}T$$
$$= 0.0074 + 1.4\times10^{-5}T$$

Heat of reaction = 12.5 + 0 -(-20.23) = 32.73 kcal/g-mol

$$\Delta H_T^0 = I_H + \Delta a T + \Delta\beta\left(\tfrac{1}{2}\right)T^2 = I_H + 0.0074T + 1.4\times10^{-5}\left(\tfrac{1}{2}\right)T^2 = 32.73$$

Then $I_H = 32.73 - 0.0074\times298 - 1.4\times10^{-5}\times\tfrac{1}{2}\times298^2 =$ 29.903 kcal/g-mol

$$\Delta G_T^0 = 29.903 - 0.0074T\ln T - \tfrac{1}{2}\left(1.4\times10^{-5}\right)T^2 + I_G(T) = 16.28 + 0 - (-7.86) = 24.14$$

or

$$\Delta G_T^0 = 29.903 - 0.0074(298)\ln 298 - \tfrac{1}{2}\left(1.4\times10^{-5}\right)(298)^2 + I_G(298) = 24.14$$

From which I_G = 0.0249

$$\Delta G_{1000}^0 = 29.903 - 0.0074T\ln T - 0.7\times10^{-5}T^2 + 0.0249T$$

$$= 29.903 - 0.0074(1000)\ln 1000 - 0.7\times10^{-5}(1000)^2 + 0.0249(1000)$$
$$= -\ \mathbf{3.31\ kcal.g\text{-}mol}$$

$$\ln K = \frac{\Delta G^0}{RT} = -\frac{-3.31\times10^3}{1.987\times(1000)} = 1.66$$ **from which K = 5.291**

Calculation of conversion

Assume the components behave as ideal gases.

$$\mathbf{K} = \frac{f_{H_2}f_{C_2H_4}}{f_{C_2H_6}} = \frac{y_{H_2}P_T y_{C_2H_4}P_T}{y_{C_2H_6}P_T} = \frac{y_{H_2}y_{C_2H_4}}{y_{C_2H_6}}$$ **since P_T** is 1 atm.

Let X mols of C_2H_6 are converted at equilibrium. Then mols of each component initially and at equilibrium are as follows (on basis of 1 mol of $\mathbf{C_2H_6}$)

Component	Initial mols	Equilibrium
C_2H_6	1.0	1 - X
C_2H_4	0.0	X
H_2	0.0	X
Total	1.0	1 + X

Therefore $y_{C_2H_6} = \frac{1-X}{1+X}$ $\quad y_{H_2} = \frac{X}{1+X}$ $\quad y_{C_2H_4} = \frac{X}{1+X}$

and $K = \frac{\left(\frac{X}{1+X}\right)\left(\frac{X}{1+X}\right)}{\frac{1-X}{1+X}} = \frac{X^2}{1-X^2} = 5.291$ and $X^2 = \frac{5.291}{1+5.291} = 0.841$

Then X = 0.917 (The positive and acceptable root of the quadratic equation.)
Therefore the conversion is 91.7 %.

13-2:

Total cost $C_T = \left(V_R - Vol.\ of reactor\right)\left(\frac{Cost}{h \cdot V_R}\right)$ + (Feed rate of reactant x cost of reactant)

For a first reaction, $V = \frac{F_{AO}X_A}{kC_{AO}(1-X_A)}$

Rate of production of R = $F_R = F_{AO}X_A$ = 250 mols/h
Now write the total cost in terms of one variable X_A eliminating F_{AO}

Total cost $C_T = \frac{F_R}{kC_{AO}(1-X_A)}C_E + \frac{F_R}{X_A}C_M$

$$= \frac{250 \times 0.02}{0.25(0.2)(1-X_A)} + \frac{250 \times 0.9}{X_A} = \frac{100}{(1-X_A)} - \frac{225}{X_A}$$

The conversion corresponding to minimum cost is found by differentiating C_T with respect to X_A and equating the result to 0. Thus

$$\frac{dC_T}{dX_A} = 0 = \frac{100}{(1-X_A)^2} - \frac{225}{X_A^2}$$

which on simplification gives $X_A^2 - 3.6X_A + 1.8 = 0$

Solving, the optimum conversion is $X_A = 0.6$.
The conditions corresponding to optimum conversion are as follows

Feed rate $F_{AO} = \frac{F_R}{X_A} = \frac{250}{0.6} = 416.7$ *mols/h*
Reactor volume $V = \frac{F_{AO}X_A}{kC_{AO}(1-X_A)} = \frac{250 \times 0.6}{0.25 \times 0.2 \times (1.0-0.6)} = 7500$ liters

Cost of product / mol product = $\frac{V(C_E)}{F_R} + \frac{F_{AO}C_M}{F_R} = \frac{7500(0.02)}{250} + \frac{416.7(0.9)}{250}$ = \$2./g-mol of product.

13-3:

For a reaction of the nth order, Half-Time is given by

$$t_{1/2} = \frac{2^{n-1}-1}{k(n-1)} \frac{1}{C_{AO}^{n-1}} \quad \text{or} \quad \ln t_{1/2} = \ln \frac{2^{n-1}}{k(n-1)} + (1-n)\ln C_{AO}$$

Thus a plot of ln C_{AO} vs ln $t_{½}$ should be a straight line . This plot for the present data is shown in Figure 13-1.
From graph, the slope of the line is - 1.
Therefore, order of reaction is given by 1 - n = -1 or n = 2 .
Order of the reaction is 2.

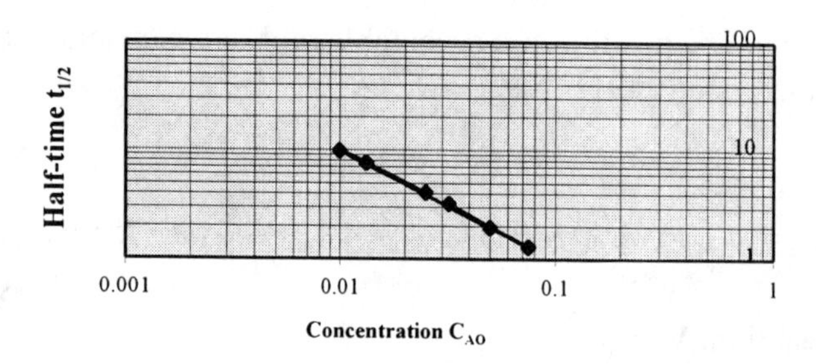

Figure 13-1. Plot of Half-times vs Concentration

Specific reaction rate at 100 ºC

$$k_{100} = \frac{2^{n-1}-1}{(n-1)} \times \frac{1}{t_{1/2}.C_{AO}^{n-1}} = \frac{2^{n-1}-1}{2-1} \times \frac{1}{5\times 0.02^{n-1}} = 10 \text{ L/ (g-mol.min)}$$

At 110 ºC, C_{AO} = 0.025 g-mol/L $t_{1/2}$ = 2 min. and $k_{110} = \frac{2-1}{2-1} \times \frac{1}{0.025\times 2} = 20$ L/(g-mol . min)

Activation energy; $\ln \frac{k_{110}}{k_{100}} = -\frac{E}{RT_2} + \frac{E}{RT_1} = -\frac{E}{R(383)} + \frac{E}{R(373)}$

or $\ln(20/10) = \frac{E}{1.987}\left[-\frac{1}{383} + \frac{1}{373}\right]$ From which E = 19676 cal/g-mol

Frequency factor

$$k = ae^{-\frac{E}{RT}} \quad \text{and} \quad a = ke^{+\frac{E}{RT}} = 10e^{+\frac{19676}{1.987\times 373}} = 3.39\times 10^{12}$$

13-4:

Since the reaction is liquid phase. volume can be considered constant. k is given at 25 °C, and reaction is carried out at 40 °C. Therefore, first calculate k at 40 °C

$$\ln\frac{k_2}{k_1} = -\frac{E(T_1 - T_2)}{RT_1T_2} = -\frac{10600(298-313)}{1.987(298)(373)} = 0.8579$$

$$\frac{k_2}{k_1} = e^{0.8579} = 2.3582$$

Hence $k_2 = k_1 \times 2.3582 = 0.158 \times 2.8579 = 0.3726$ min^{-1}

The rate equation is therefore $-r_A = k_2 C_{A0}(1 - X_A)$ at 40 °C

Calculate anhydride fed to the reactor.
Acetic acid produced = 300 kg/h = 300/60 = 5 kmols/h
2 mols of CH_3COOH require hydrolysis of 1 mol of acetic anhydride and the conversion is 95 %
Therefore, anhydride needed = 5 x (1/2) x (1/0.95) = 2.632 kmol/h = 0.0439 kmol/min.

For plug-flow reactor,

$$V = -\frac{F_{AO}}{C_{AO}}\int_0^{0.95}\frac{dX_A}{k(1-X_A)} = -\frac{0.0439}{0.1\times10^{-3}}\int_0^{0.95}\frac{dX_A}{0.373(1-X_A)} = -1176.9\int_0^{0.95}\frac{dX_A}{1-X_A} = 1176.9\left[\frac{1}{1-X_A}\right]_0^{0.95}$$

$$= 1176.9\left(\ln\frac{1}{1-0.95}\right) = 352 \text{ L}$$

13-5:

For a reaction taking place at constant volume , the time t is given by

$$t = -\int_{C_{AO}}^{C_A}\frac{dC_A}{-r_A}$$

A plot of $1/-r_A$ vs C_A will allow to determine the time of reaction for a given conversion.

The values of ($1/-r_A$) are tabulated below and plotted in figure 13-2.

C_A (g-mol/L)	0.1	0.2	0.3	0.4	0.5	0.6	0.7	1.3	2.0
$-r_A$ g-mol/L.min	0.1	0.3	0.5	0.6	0.5	0.25	0.1	0.045	0.04
$1/-r_A$	1.0	3.34	2.0	1.67	2.0	4.0	10	16.67	20.0

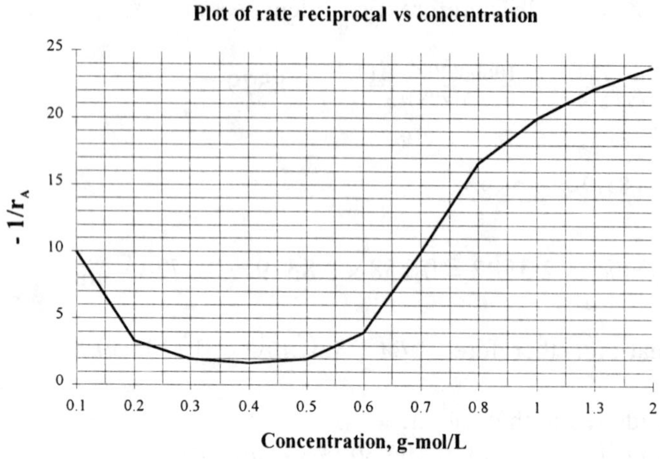

Plot of rate reciprocal vs concentration

- 1/r_A

0 5 10 15 20 25

0.1 0.2 0.3 0.4 0.5 0.6 0.7 0.8 1 1.3 2

Concentration, g-mol/L

Figure 13-2:A plot of $(-1/r_A)$ vs concentration C_A

Obtain the value of the integral between $C_A = 0.3$ and $C_{AO} = 1.3$ g-mol/L. Use Simpson's rule to get the area under the curve. The area is calculated as

Area under the curve =

$$-\frac{0.1}{3}[2 + 22.22 + (1.67 + 4 + 16.67 + 20 + 21.7) + 2(2 + 10 + 20 + 20.9)$$

$$= -(-12.9) = 12.9 \text{ min.}$$

(b) The difference between the measured rate and calculated rate can be expressed as

$$e_u = (-r_A)_u - kC_{Au}$$

Sum of the squares of the errors is given by

$$\Sigma_{u=1}^{n} e_u^2 = \Sigma[(-r_A) - kC_A]^2$$

Value of k that minimizes the error function is required . For this we differentiate with respect to k and get the differential as follows

$$\frac{\delta}{\delta k}[\Sigma e_u^2] = \frac{\delta}{\delta k}\Sigma[(-r_A)_u - kC_{Au}]^2 = \frac{\delta}{\delta k}\Sigma[r_{Au}^2 + 2r_{Au}kC_{Au} + k^2C_{Au}^2] = \Sigma[0 + 2r_{Au}C_{Au} + 2kC_{Au}^2]$$

or $$\Sigma_{u-1}^{n}(-r_A)\, C_{Au} = \bar{k}\,\Sigma C_A^2$$

The values of $-r_{Au}C_{Au}$ and C_{Au}^2 are listed in table below:

$-r_A C_{Au}$	0.01	0.06	0.15	0.24	0.25	0.15	0.07	0.048	0.05	0.0585	0.084
C_A^2	0.01	0.04	0.09	0.16	0.25	0.36	0.49	0.64	1.00	1.69	4.0

$\Sigma -r_A C_{Au} = 1.1705$ $\qquad \Sigma C_{Au}^2 = 8.73$

Then $k = \frac{1.1705}{8.73} = 0.134$ min^{-1} and therefore rate equation is $-r_A = 0.134 C_A$

Process Control

14-1: (a) From first principles, develop transfer functions and a block diagram for the control system shown in figure below. Assume that the transfer function for the measurement dynamic lag is given by

$$\frac{\theta_m(s)}{\theta(s)} = \frac{1}{\tau_m s + 1}$$

Where θ and θ_m are temperature deviations and a proportional control is used.
$\theta = T_i - T_s$; $\theta_m = T_m - T_{ms}$ where T = temperature. subscript i indicates transient value and s indicates steady state value.

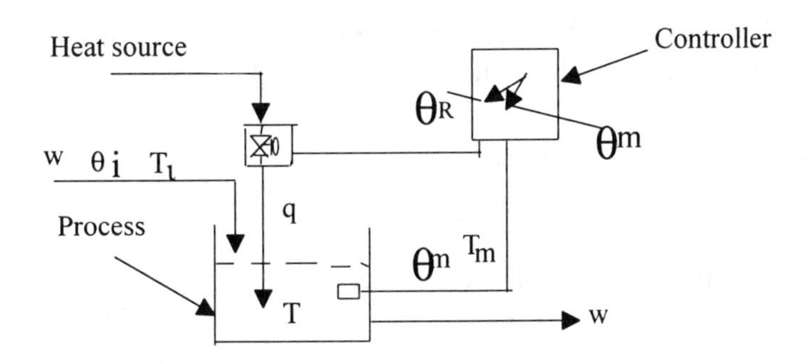

(b) Derive expressions for the period of oscillation and the damping coefficient ξ in terms of the time constants and controller gain K_c assuming proportional control and regulatory operation.

14-2: A control system with a derivative control mode added is shown in the figure below. It has a first order process and a first order measurement lag.

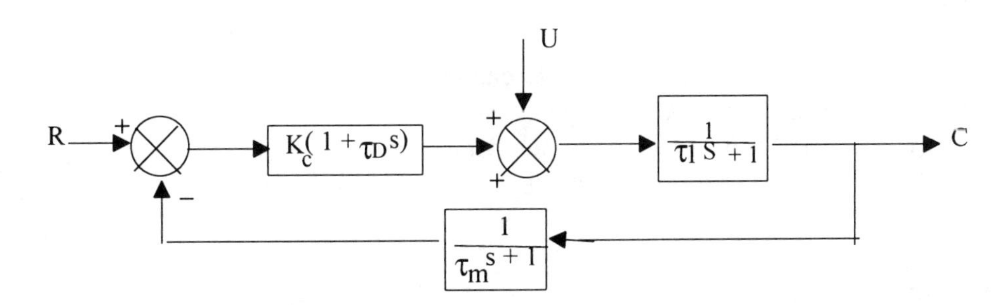

The characteristic equation for the control system is as follows.

$$\frac{\tau_1 \tau_m}{1+K} s^2 + \frac{\tau_1 + \tau_m + K\tau_D}{1+K} s + 1 = 0$$

Assume close loop regulator operation. and $\tau_1 = 2$ min, $\tau_m = 0.25$ m, and $\tau_D = 3$ s and $K_c = 3$
Answer the following questions by making the correct choice.

14-2.1 The period of oscillation in minutes is very nearly
a) 0.25 b) 2.5 c) 0.5 d) 0.7

14-2.2 The value of the damping coefficient is very nearly
a) 0.35 b) 0.7 c) 1.0 d) 0

14-2.3 The overshoot is very nearly
a) 0.092 b) 0.011 c) 0.046 d) 0.20

14-2.4 The decay ratio is very nearly
a) 0.002116 b) 0.092 c) 0.46 d) 0.046

14-2.5 If $\tau_D = 0$, the value of K_c required for critical damping is very nearly
a) 0.5625 b) 5.625 c) 56.3 d) 0.75

14-2.6 The radian frequency in radians/s is very nearly
a) 2.857 b) 4.0 c) 0.714 d) 0.0476

14-2.7 The cyclical frequency (cycles/min)is very nearly
a) 4.55 b) 0.4546 c) 0.046 d) 0.092

14-2.8 The ratio of actual frequency to natural frequency is very nearly
a) 0.3 b) 0.51 c) 0.714 d) 0.95

14-2.9 If an integral control were added to the system and a step change of one unit was given to load, the steady state offset will be very nearly
a) 0 b) 0.25 c) 0.10 d) 0.025

14-2.10 Which of the following statements correctly describes the stability of the system?

(A) System is oscillatory or underdamped
(B) System is critically damped
(C) System is overdamped or non-oscillatory
(D) System is inherently unstable.

Solutions to the Problems

14-1:

Take an unsteady state energy balance for the tank system. This can be written as follows

$$\rho c V \frac{dT}{dt} = wC(T_i - T_r) - wC(T - T_r) + q(t)$$

where ρ = density of the fluid
C = specific heat of the fluid
w = flow of fluid
T = average temperature of the fluid in the tank at any time t
T_i = temperature of fluid flowing into the tank at time t
q(t) = heat input to the system
T_r = reference temperature

In writing the above balance, usual assumptions of constancy of variables ρ, C, and w is made. Also the fluid in the tank is assumed well mixed at all times so that the bulk temperature of the fluid is the same throughout in the tank and the temperature of fluid leaving the system is the same as that of the bulk temperature. Also w in = w out assumed.

For steady state energy balance, the accumulation term is zero i.e. $\frac{dT}{dt} = 0$ and it can be written as

$$0 = wC\left(T_{is} - T_r\right) - wc(T_s - T_r) + q_s$$

Now subtract the steady state equation from the transient equation and get the following differential equation

$$\rho CV\frac{d(T-T_s)}{dt} = wC\left[\left(T_i - T_{is}\right) - (T - T_s)\right] + q(t) - q_s$$

In the above equations, subscript s denotes steady state.
(The reference temperature does not appear in the final equation because it cancels out.)

Now make the following substitutions and get the deviation variables which also simplifies the writing of the equation.

$$\theta_i = T_i - T_{is} \qquad \theta = T - T_s \qquad Q = q(t) - q_s$$

Then the deviation differential equation becomes

$$\rho CV\frac{d\theta}{dt} = wC\left(\theta_i - \theta\right) + Q$$

We now take Laplace transform of the equation to obtain

$$\rho CVs\theta(s) = wC[\theta_i(s) - \theta(s)] + Q(s)$$

By rearrangement, the equation is written as follows

$$\theta(s)\left(\frac{\rho V}{w}s + 1\right) = \frac{1}{wC}Q(s) + \theta_i(s)$$

which can be written as an expression for $\theta(s)$ in the following manner

$$\theta(s) = \frac{\frac{1}{wC}}{\tau s + 1}Q(s) + \frac{1}{\tau s + 1}\theta_i(s)$$

where $\tau = \frac{\rho V}{w}$ is the time constant

If there is a change only in Q(t), $\theta_i(t) = 0$. Therefore transfer function relating θ to Q is

$$\frac{\theta(s)}{Q(s)} = \frac{1/wC}{\tau s + 1}$$

Similarly if there is a change in $\theta_i(t)$ only, Q(t) = 0. Then transfer function relating θ to θ_i is as follows

$$\frac{\theta(s)}{\theta_i(s)} = \frac{1}{\tau s + 1}$$

The block diagram for the process can therefore be drawn in figure (a) below. In figure b is given equivalent representation.

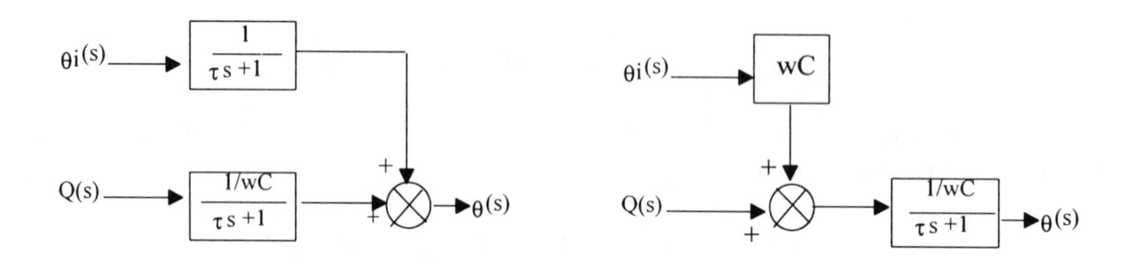

(a) (b)

Block diagram for process

The block diagram for the control system is given below.

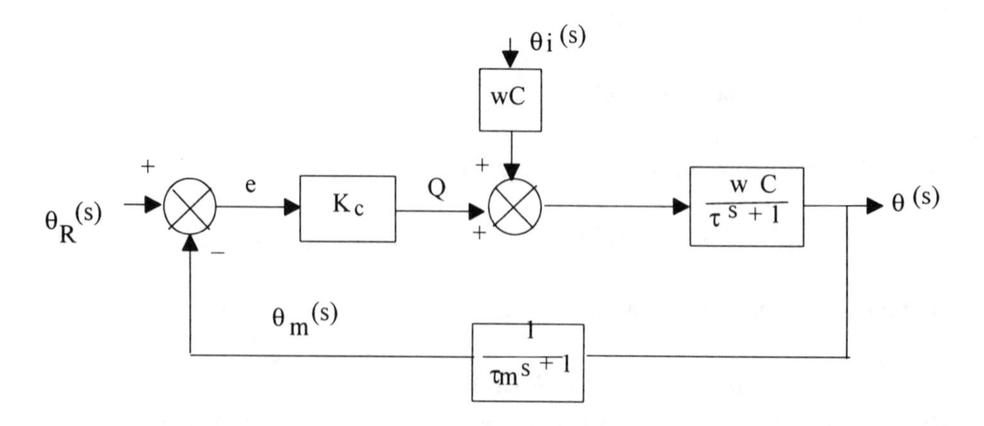

In the above diagram , the output variable is measured and compared with the set point represented by $\theta_R(s)$. $\frac{1}{\tau_m s + 1}$ represents the transfer function for the time lag in the measurement . K_C represents the controller gain

(b) The characteristic equation is

$$\left(\tau_1 s+1\right)(\tau_m s+1)+w^2C^2K_c=0$$

Simplification gives $\tau_1\tau_m s^2+\left(\tau_1+\tau_m\right)s+1 + w^2C^2K_c=0$

Dividing by $1+ w^2C^2K_c$, to put the equation in standard form

$$\frac{\tau_1\tau_m}{1+ w^2C^2K_c}s^2 + \frac{\tau_1+\tau_m}{1+ w^2C^2K_C} + 1=0$$

$$\tau^2 = \frac{\tau_1\tau_m}{1+w^2C^2K_c} \quad \text{then} \quad \tau = \sqrt{\frac{\tau_1\tau_m}{1+w^2C^2K_c}}$$

and $$2\xi\tau = \frac{\tau_1+\tau_m}{1+w^2C^2K_c}$$

$$\xi = \frac{\tau_1+\tau_m}{1+w^2C^2K_c}\frac{1}{2\tau} = \frac{1}{2\sqrt{\frac{\tau_1\tau_m}{1+w^2C^2K_c}}}$$

$$\xi = \frac{\tau_1+\tau_m}{2\sqrt{\tau_1\tau_m}}\frac{1}{\sqrt{1+w^2C^2K_c}}$$

14-2:

14-2.1 $$\tau = \sqrt{\frac{\tau_m\tau_1}{1+K_c}} = \sqrt{\frac{0.25\times1}{1+3}} = 0.25 \text{ min.}$$

Answer is c)

14-2.2 $$\xi = \frac{\tau_1+\tau_m+K_c\tau_D}{1+K} \times \frac{1}{2}\sqrt{\frac{1+K_c}{\tau_1\tau_m}}$$

$$= \frac{1+ 0.25+3\times0.0}{1+3} = 0.7$$

Answer is b)

14-2.3 Overshoot $= \exp -\left(\pi\xi\right)/\sqrt{1-\xi^2}$

$= \exp -\pi(0.7)/\sqrt{1-0.7^2} = 0.046$

Answer is a)

14-2.4 The decay ratio is equal to square of the overshoot
Therefore decay ratio = $(0.046)^2 = 0.002116$

Answer is d)

14-2.5 For critical damping, $\xi = 1.0$ Substituting in relation for ξ and for $\tau_D = 0$,

$$1 = \frac{0.25+1.0+ 0.\times K_c}{1+K_c} \times \frac{1}{2}\sqrt{\frac{1+K_c}{0.25\times1}}$$

Solving for Kc = 0.5625

Answer is a)

14-2.6 $\omega = \frac{\sqrt{1-\xi^2}}{\tau} = \frac{\sqrt{1-0.7^2}}{0.25} = 2.857$ radian/min – **0.04762** radian/sec

Answer is d)

14-2.7 Cyclical frequency $= f = \frac{1}{2\pi}\frac{\sqrt{1-\xi^2}}{\tau} = \frac{1}{2\pi}\frac{\sqrt{1-0.7^2}}{0.25} = 0.4546$ cycles/min

Cyclical frequency in cycles/min = 0.4546

Answer is b)

14-2.8 Natural frequency is related to actual frequency by the relation

$$\frac{f}{f_n} = \sqrt{1-\xi^2} = \sqrt{1-0.7^2} = 0.714$$

Answer is c)

14-2.9 Integral action is added to the proportional controller to eliminate offset
Therefore offset is equal to zero.

Answer is a)

14-2.10 A second order system is inherently stable. In this case $\xi = 0.7 < 1.0$
Therefore based on the value of ξ, the system is underdamped and oscillatory.

Answer is a)

MASS TRANSFER

15-1: Estimate the rate of absorption of SO_2 in a water film flowing down a vertical wall 1.2 m long at the rate of 252 kg/h. per meter of width at 30 °C and 1 atm. pressure at which the solubility of SO_2 is 7.81 g/100 g of water. The gas is pure SO_2 at 1 atm. Solution density may be taken as 990 kg/m³, viscosity of solution = 0.8 cP at 30 °C and 0.654 cP at 40 °C. Diffusivity of SO_2 in water = 2.6 x10⁻⁵ cm²/s at 40 °C.

15-2: A laboratory wetted-wall column 2" ID and 6' in length was used to absorb SO_2 in the falling film of water at 30 °C. and 1 atm. pressure. The water rate was maintained at 500 lb/h.Pure SO_2 at 30 °C and 1 atm is fed to the column . The other data are as follows : density of solution = 61.8 lb/ft³

Diffusivities of SO_2 at 30 °C: (a) in air = 0.553 ft²/h. (2) in water = 7.56x10⁻⁵ ft²/h.

Solubility of SO_2 = 7.81 g/100 g of water. viscosity of solution = 0.8 cP.

A plot of equilibrium relationship for SO_2-H_2O system showing partial pressure in atm vs concentration in lb-mol/ft³ is given in Figure 15-1.

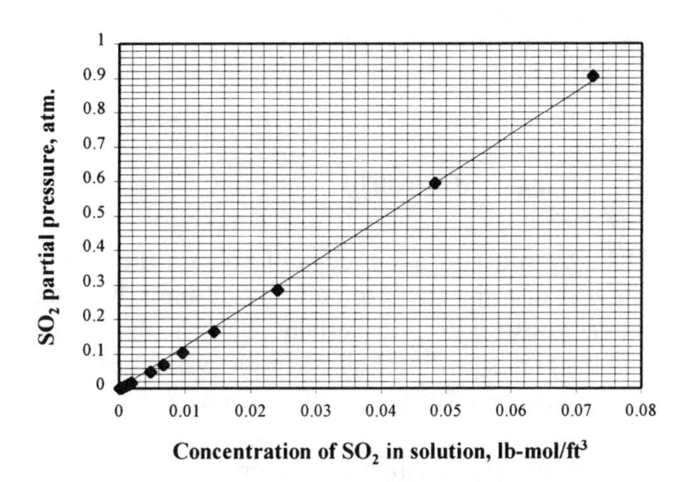

Figure 15-1: SO² partial pressure vs concentration.

Answer the following questions by selecting the correct answer.

15-2.1 The Reynolds' number for the falling film is very nearly

a) 15.3 b) 7623 c) 1973 d) 54780

15-2.2 The thickness of the falling film in ft is very nearly

a) 1.5×10^{-3} b) 1.11×10^{-3} c) 1.2×10^{-3} d) 1.35×10^{-3}

15-2.3 The maximum velocity of the falling film in ft/s is very nearly

a) 2.86 b) 3.4 c) 4.3 d) 2.3

15-2.4 The value of Peclet number is very nearly
a) 1973 b) 14764 c) 40145 d) 817565

15-2.5 The value of the liquid phase mass transfer coefficient in lb-mol/(h.ft^2. lb-mol/ft^3) is very nearly
a) 1.656 b) 0.49 c) 32.9 d) 7.5

15-2.6 Interphase concentration of SO_2 in liquid phase in units of lb-mol/(h.ft^2.lb-mol/ft^3) is very nearly
a) 0.0762 b) 0.007 c) 0.07 d) 0.00762

15-2.7 The total absorption rate in mols /h is very nearly
a) 1.43×10^{-3} b) 0.27 c) 2.7 d) 7.48×10^{-5}

15-2.8 The gas phase average mass transfer coefficient k_G in lb-mol/(h.ft^2.atm.) is very nearly
a) 0.633 b) 0.27 c) 3.798 d) 1.99

15-2.9 The average overall mass transfer coefficient K_G in lb-mols/(h.ft^2.atm) based on gas phase is
a) 1.58 b) 0.767 c) 0.1106 d) 2.0

15-2.10 The average overall mass transfer coefficient K_L in lb-mols/(h.ft^2.lb-mol/ft^3) based on gas phase is
a) 1.37 b) 1.3 c) 0.46 d) 0.603

15-3: In absorption of NH_3 by water in a wetted-wall column the overall mass transfer coefficient based on gas phase was determined at to be 0.1185 lb-mol/(h.ft^{2}atm). The gas phase resistance is 98.7% of the total resistance. At a certain point, in the column a gas sample analyzed 20% by volume and the corresponding liquid sample was found to contain 5 % NH_3. Column instruments show that absorption takes place at 1 atm and 68 °F.

(a) Estimate the individual coefficients of mass transfer in each phase.
(b) Estimate the composition of NH_3 at the interface.
(c) Express the overall mass transfer coefficient based on liquid phase.
(d) convert k_y into k_G and k_C

Equilibrium relationship for NH_3-H_2O system are plotted in Figure 15-2.
Diffusivity of ammonia in air at 68 °F and 1 atm = 0.853 ft^2/h
Diffusivity of NH_3 in water at 68 °F = 10.1×10^{-5} ft^2/h

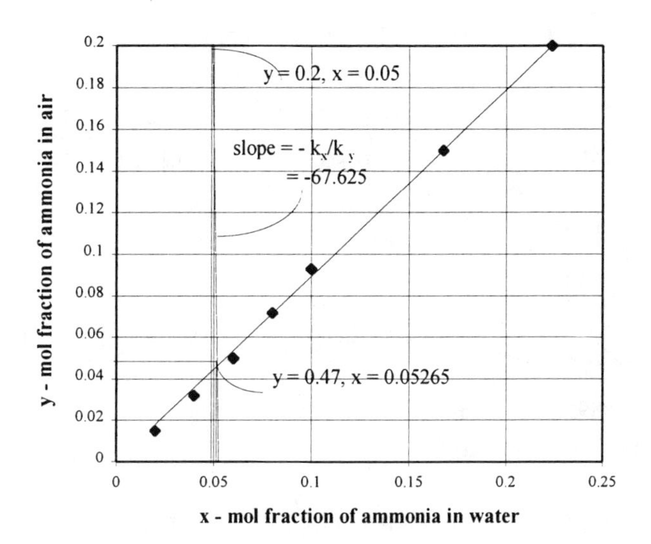

Figure 15-2. y vs x for NH³-H₂O system

Solutions to problems:

15-1:

Solubility of SO_2 at 30 °C. and 1 atm. = $[(7.81/64)/100] \times 1000 = 1.22$ g-mol/liter
= 1.22 kmol/m³

Therefore, concentration at the interface = C_{Ai} = 1.22 kmol/m³ , $\mu = 0.8$ cP
= $8.0\text{x}10^{-4}$ kg/m s ,

L = 1.2 m. Mass flow rate = 252/3600 = 0.07 kg/s per meter width.

Calculate film thickness $\delta = \left(\frac{3\mu\Gamma}{\rho^2 g}\right)^{1/3} = \left(\frac{3\times8\times10^{-4}\times0.07}{990^2\times9.81}\right)^{1/3} = 2.55\times10^{-4}\ m$

Reynolds' number $\text{Re} = \frac{4\Gamma}{\mu} = \frac{4(0.07)}{8\times10^{-4}} = 350$

This is greater than 100. and solution for large Reynolds' numbers or short contact time is to be used.

Diffusivity of SO_2 at 30 °C = $\frac{303}{313} \times \frac{0.653}{0.8} \times 2.6\times10^{-} = 2.05\times10^{-9}$ m²/s.
(By Wilke-Chang equation).

Thus, $k_{L,av} = \left(\frac{6D_{AB}\Gamma}{\pi\rho\delta L}\right)^{1/2} = \left[\frac{6(2.05\times10^{-9})(0.07)}{\pi\times990(2.55\times10^{-4})1.2}\right]^{1/2} = 3.008\times10^{-5}\ kmol/m^{2\cdot}\cdot s\cdot\left(kmol/m^3\right)$

At the top of the plate, $C_{Ai} - C_{Ao} = C_{Ai} = 1.22\ kmol/m^3$

At the bottom of the plate , $C_{Ai} - C_{A,L} = 1.22 - \bar{C}_{A,L}$

Average velocity $\bar{u}_y = \frac{\Gamma}{\rho\delta} = \frac{0.07}{990\times(2.55\times10^{-4})} = 0.277 \quad m/$

Flux is given by $N_{A,av} = \frac{\bar{u}\delta}{L}\left(c_{\bar{A},L} - C_{Ao}\right) = k_{L,av}(C_{A,i} - C_{A,})_M$

Now $$\left(C_{A,i} - C_A\right)_M = \frac{\left(C_{A,i} - C_{A,o}\right) - \left(C_{A,l} - C_{A,i}\right)}{\ln\left[\left(C_{A,i} - C_{Ao}\right)/\left(C_{A,i} - C_{A,L}\right)\right]} = \frac{1.22 - \left(1.22 - C_{A,L}\right)}{\ln\left[1.22/\left(1.22 - C_{A,L}\right)\right]}$$

Therefore $$\frac{\bar{u}\delta}{L}\left(C_{\bar{A},L}\right) = \frac{k_{L,av}\left[1.22 - \left(1.22 - C_{A,L}\right)\right]}{\ln\left[1.22/\left(1.22 - C_{A,L}\right)\right]}$$

Substitution of known values gives $$\frac{0.277(2.55\times10^{-4})C_{A,L}}{1.2} = \frac{(3.008\times10^{-5})\left[1.22 - \left(1.22 - C_{A,L}\right)\right]}{\ln\left[1.22/\left(1.22 - C_{A,L}\right)\right]}$$

From which $C_{A,L} = 0.4895$ kmol/m³.
Then rate of absorption can be estimated as

$N_{Az} = \bar{u}_y\,\delta(C_{A,L} - C_{Ao})/1.2 = 0.277\left(2.55\times10^{-4}\right)(0.4895 - 0)/1.2 = 2.88\times10^{-5}$ kmol/ per unit width

15-2:

15-2.1 $\Gamma = \frac{500}{\pi D} = \frac{500}{\pi(2/12)} = 95$ lb/h.ft

$$Re = \frac{4\Gamma}{\mu} = \frac{4\times955}{0.8\times2.42} = 1973$$

Answer is c)

15-2.2 $\delta = \left(\frac{3\mu\Gamma}{\rho^2 g}\right)^{1/3} = \left(\frac{3\times0.8\times2.42\times955}{61.8^2\times4.18\times10^8}\right)^{1/3} = 1.5\times10^{-3}$ ft

Answer is a)

15-2.3 Average velocity $\bar{u}_y = \frac{\Gamma}{\rho\delta\times3600} = \frac{955}{61.8\times1.5\times10^{-3}\times3600} = 2.86$ ft/s

Maximum velocity $= 1.5\ u_y = 1.5\times2.862 = 4.3$ ft/s

Answer is c)

15-2.4 Re = 1973 $\quad$ $Sc = \frac{\mu}{\rho D_{AB}} = \frac{0.8\times2.42}{61.8\times7.56\times10^{-}} = 414.4$

Peclet number = Pe = Re x Sc = 1973 x 414.4 = 817611

Answer is d)

15-2.5 Sherewood number, $Sh = 1.76\times10^{-5}(Re)^{1.506}(Sc)^{0.5}$ since $1300 < Re = 1973 < 8300$

$$= 1.76\times10^{-5}(1973)^{1.506}(414.4)^{0.5} = 32.86$$

$$\frac{k_{L,av}\delta}{D_{AB}} = 32.86$$

$$k_{L,av} = 32.86 \times \frac{D_{AB}}{\delta} = 32.86 \times \frac{7.56\times10^{-5}}{1.5\times10^{-3}} = 1.656 \text{ ft/h or lb-mol/(h.ft}^2\text{.lb-mol/ft}^3)$$

Answer is a)

15-2.6 Solubility of SO_2 in water at 30 °C = 7.81 g/100g of water
specific gravity of solution = 0.99
volume of solution when 7.81 g of SO_2 are dissolved in 100 g of water = $\frac{100+7.81}{0.99} = 108.9$ cc

then concentration of $SO_2 = \frac{(7.81/64)}{108.9} \times 1000$ g-mol/L

= 1.1206 g-mol/L = 1.1206 kmol/m^3

= $1.1206\frac{2.2046}{35.3} = 0.07$ lb-mol/ft^3

Answer is c)

15-2.7 Bulk concentration at top $C_L = 0$ (feed water contains no SO_2)

$$\frac{2.862\times1.5\times10^{-3}C_{AL}}{6} = \frac{(1.656/3600)C_{AI}}{\ln\frac{0.07}{0.07 - C_{AL}}}$$

Solving for C_{AL}, gives $C_{AL} = 0.0332$ lb-mol/(h.ft^2.lb-mol/ft^3)

Absorption rate = (2.862)(1.5x10^{-3}) (0.0332 - 0) ($\pi\times0.167$)(3600) = 0.27 lb-mol/h

Answer is b)

15-2.8 Interphase SO_2 concentration $C_{A,i}$ = 0.07 lb-mol/ft^3
From equilibrium line, slope H = 12.35 atm/(lb-mol/ft^3)
Therefore $P^* = HC_{A,i}$ = 12.35 (0.07) = 0.8645 atm.
Driving force = P - P* = 1 - 0.8645 = 0.1355 atm.
Then $N_A = k_G A.(P - P^*) = 0.27$ lb-mol/h
$k_G(\pi\times0.167\times6)(1 - 0.8645) = 0.27$
From which k_G = 0.633 lb-mol/(h.ft^2.atm)

Answer is a)

15-2.9 $\frac{1}{K_G} = \frac{1}{k_G} + \frac{H}{k_L} = \frac{1}{0.633} + \frac{12.35}{1.656} = 9.0375$

K_G = 0.1106 lb-mol/(h.ft^2 atm.)

Answer is c)

15-2.10 $\frac{1}{K_L} = \frac{1}{k_L} + \frac{1}{Hk_G} = \frac{1}{1.656} + \frac{1}{12.35\times0.63} = 0.73178$

K_L = 1.37 lb-mol/(h.ft^2.lb-mol/ft^3)

Answer is a)

15-3:

Gas and liquid phase mass transfer coefficients:
Resistance to mass transfer in gas phase = (1/K_y) x 0.98 = (1/0.1185)(0.987) = 8.33
Resistance to mass transfer in liquid phase = (1/.1185)(.013) = 0.11

$1/k_y = 8.33$ therefore $k_y = 0.12$ lb-mol/(h.ft^2.mol fraction.)

Similarly $\frac{0.8927}{k_x} = 0.11$

Therefore $k_x = \frac{0.8927}{0.11} = 8.115$ lb-mol/(h..ft^2.mol fraction)

Interface compositions:

Locate y = 0.2 , x = 0.05 on the y-x diagram

Draw a straight line with slope $= -\frac{k_x}{k_y}$

$= -\frac{8.115}{0.12} = -67.625$

This is shown in Figure Mass 2. However in this case, it is preferable to obtain the concentrations from the equation for the slope using arithmetic as follows:

Since the point denoted by interfacial compositions, lies on both the slope line and on the equilibrium line, $y_{AS} = 0.8927\, x_{AS}$

$$-\frac{k_x}{k_y} = \frac{y_A - y_{AS}}{x_A - x_{AS}} = -67.625 = \frac{0.2 - y_{AS}}{0.05 - x_{AS}} = \frac{0.2 - 0.8927 x_{AS}}{0.05 - x_{AS}}$$

from which $x_{AS} = 0.0523$ Then $y_{AS} = 0.8927(0.523) = 0.047$

Overall mass transfer coefficient based on liquid phase concentrations:

$$\frac{1}{K_x} = \frac{1}{k_x} + \frac{1}{0.8927 k_G} = \frac{1}{8.115} + \frac{1}{0.8927 \times 0.633} = 1.8$$

Therefore $K_x = 0.5283$ lb-mol/(h.ft^2 mol fraction)

Conversion of k_y into k_G and k_C for gas phase:

$\mathbf{k_G} = \mathbf{k_y/p_t} = \mathbf{k_C/RT}$

Since $p_t = 1$ atm , $\mathbf{k_G} = k_y$ lb-mol/(h.ft^2.atm) = 0.12 lb-mol/(h.ft^2.atm)

$\mathbf{k_C} = \mathbf{k_G}$ RT = 0.12 (0.73) (68 + 460) = 46.3 ft/h = 46.3 lb-mol/(h.ft^2.lb-mol/ft^3)

PLANT SAFETY

16-1

The lower and upper flammability limits(LFL and UFL) of methane and hexane in air are as follows.

	LFL % v/v	UFL % v/v
Methane(CH_4)	5.0	15
Hexane(C_6H_{14})	1.1	7.5

The following is the composition of a gas mixture containing methane, hexane, and air in a reactor:

	% v/v
Methane	12
Hexane	8
Air	80

16-1.1

In the combustibles, i.e., on an air-free basis, the % (v/v) of methane is
(A)12 (B)15 (C)60 (D)0.6

16-1.2

In the combustibles, the %(v/v) of hexane is
(A)88 (B)85 (C)40 (D)99.4

16-1.3

The lower flammability(LFL) limit of the mixture in air is
(A)2.07% (B)3.59% (C)1.45% (D)2.56%

16-1.4

The upper flammability(UFL) limit of the mixture in air is
(A)20.4% (B)5.8% (C)15.37% (D)10.71%

16-1.5

The stoichiometric coefficient of oxygen in the combustion reaction in which one mole of the combustibles comprising methane and hexane is completely burnt to produce carbon dioxide and water is
(A)12 (B) 0.5 (C)5 (D)3.2

16-1.6

The stoichiometric concentration of combustibles in %(v/v) is
(A)12% (B)4.03% (C)10.09% (D)20%

16-1.7

The estimated LFL from stoichiometric concentration of combustibles(LFL = $0.55C_{st}$) is:
(A)2.22% (B)1.5% (C) 4.03% (D)3.5%

16-1.8

The estimated UFL from stoichiometric concentration of combustibles(UFL = $3.5C_{st}$) is:
(A)6.29% (B)80.45% (C)41.03% (D)14.11%

16-1.9

The estimated time in minutes to lower the concentration of combustibles to 0.5% (v/v) from a container volume of 10,000 cuft by sweeping an inert at 1000 cfm, assuming the non-ideal mixing factor of 0.2, is:
(A)37 minutes (B)184 minutes (C) 301 minutes (D)4 minutes

16-1.10

The estimated number of purging cycles required to lower the concentration of the combustibles to 0.5%(v/v) by pressurizing with an inert to 50 psig followed by venting to atmospheric pressure, assuming ideal mixing, is:
(A)3 (B)10 (C)15 (D)20

16-2

A vertical cylindrical tank containing toluene is to be protected by a conventional relief valve for emergency relief in case of exposure to fire. The following data are available.

Design pressure:	100 psig
Relief valve set pressure:	100 psig
Elevation of bottom tangent line from grade:	6 ft
Diameter(inside):	8 ft
Height(tangent to tangent):	15 ft
Top and bottom head type:	2:1 ellipsoidal
Height of liquid level from bottom tang. Line:	15 ft
Capacity of each head:	501.3 gallons
Insulation(fire proof) thickness:	2 inch
Governing code:	OSHA 1910.106

Assume all vapor flow.
C_p / C_v of toluene at relief condition is 1.0839. Compressibility factor of toluene at relief condition is 0.831.

The vapor pressure - temperature relationship of toluene is given by:

$$\ln p_{mm\,HgA} = 16.266 - \frac{3242.4}{T - 47.181}$$

$$L = 230.33\left(1 - \frac{T}{591.7}\right)^{0.38}$$

Where L = latent heat of toluene, Btu/lb, and T = temperature in K. Calculate the relief area, and select the size of a relief valve.

16-3

After you sized the relief valve as in example 16-2 above, the production engineer informed you that the toluene is a little dirty and tends to foam when boiled. Suspecting two-phase flow, size the relief valve for two-phase flow using the following Leung equations:

$$G_T = 0.9\psi\frac{\Delta H_V}{v_{fg}}\sqrt{\frac{g_c}{C_p T_s}}$$

$$A = \frac{Q m_0 V_{fg}}{G_T V(\Delta H_v)}$$

Where:
A = relief area, m^2
C_p = specific heat at constant pressure, J/kg.K , = 2509.6 J/kg.K for this problem.
g_c = 1 $(kg.m/s^2)/N$
G_T = relief mass flux, $kg/m^2.s$
ΔH_v = latent heat of vaporization, J/kg = 2.826×10^5 J/kg for this problem
m_0 = initial mass of liquid in vessel, kg
Q = constant heat input, J/s
T_s = temperature corresponding to set pressure, K
V = gross volume of the vessel, m^3

16-4

v_{fg} = difference of specific volumes of vapor and liquid, m^3/kg = 0.03734 m^3/kg
ψ = correction factor dependent on the L/D ratio of relief line.
Liquid density at initial condition = 655.2067 kg/m^3 for this problem.

SOLUTIONS

16-1
16-1.1

Methane in the combustibles = 12/(12 + 8) = 0.6 = 60%, **Answer: C**

16-1.2

Hexane in the combustibles = 8/(12 +8) = 40%, **Answer : C**

16-1.3

$$LFL_{mixture} = \frac{1}{\frac{y_1}{LFL_1} + \frac{y_2}{LFL_2}} = \frac{1}{\frac{0.6}{5} + \frac{0.4}{1.1}} = 2.07\%$$

Answer: A

16-1.4

$$UFL_{mixture} = \frac{1}{\frac{y_1}{UFL_1} + \frac{y_2}{UFL_2}} = \frac{1}{\frac{0.6}{15} + \frac{0.4}{7.5}} = 10.71\%$$

Answer: D

16-1.5

One mole of combustibles contains 0.6 mole of methane and 0.4 mole of hexane. The stoichiometric oxygen for each of these components may be calculated as follows.

16-6

$$0.6\,CH_4 + 1.2\,O_2 \rightarrow 0.6\,CO_2 + 1.2\,H_2O$$

$$0.4\,C_6H_{14} + 3.8\,O_2 \rightarrow 2.4\,CO_2 + 2.8\,H_2O$$

Therefore, stoichiometric moles of oxygen = 1.2 + 3.8 = 5 moles oxygen per mole of fuel.

Answer: C

16-1.6

C_{st} = stoichiometric concentration of combustibles in %(v/v) may be estimated by:

$$C_{st} = \frac{100}{\dfrac{stoichiometric\ moles\ of\ oxygen\ per\ mole\ fuel}{0.21} + 1} = \frac{100}{\dfrac{5}{0.21} + 1} = 4.03\%$$

Alternatively, C_{st} may be calculated from stoichiometric material balance. Thus:

Fuel =	1.00 mole
Oxygen = (1.2 + 3.8) =	5.00 moles
Nitrogen = 5(79/21) =	18.81 moles
Total =	24.81 moles

C_{st} = 1/24.81 = 4.03%

Answer: B

16-1.7

LFL = 0.55C_{st} = 0.55(4.03) = 2.22% **Answer: A**

16-1.8

UFL = 3.5C_{st} = 3.5(4.03) = 14.11% **Answer: D**

16-1.9

$$t = \frac{V}{KQ}\ln\left(\frac{C_0}{C}\right) = \frac{10000}{0.2\times 1000}\ln\left(\frac{20}{0.5}\right) = 184.4 \ minutes$$

Answer: B

16-1.10

$$n = \frac{\ln\left(\frac{y_0}{y_n}\right)}{\ln\left(\frac{P_h}{P_l}\right)} = \frac{\ln\left(\frac{20}{0.5}\right)}{\ln\left(\frac{50 + 14.696}{14.696}\right)} = 2.5$$

Answer: A

16-2

Step 1. Calculate the exposed area.

The elevation of the top of top head = elevation of bottom tangent line + T-T height + depth of head

= 6 + 15 + 2 = 23 ft < 30 ft covered by OSHA

Therefore, the entire area of the tank is to considered for fire exposure.

$$The\ exposed\ area = 2\times 1.19\times 8^2 + \pi\times 8\times 15 = 529.31\ ft^2$$

Step 2. Calculate the fire heat input.

$$Q = 199300FA^{0.566} = 199300\times 0.3\times (529.31)^{0.566} = 2.081\times 10^6\ Btu/h$$

Step 3. Calculate the relief pressure

Relief pressure = Set pressure, psig(1 + % over pressure) + 14.7 psia
=100(1+0.21) + 14.7 = 135.7 psia = 7015.78 mmHgA

Step 4. Calculate the relief temperature

$$T = 47.181 + \frac{3242.4}{16.266 - \ln p_{mmHgA}} = 47.181 + \frac{3242.4}{16.266 - \ln(7015.78)}$$

$$= 484.75 \text{ K}$$

Step 5. Calculate latent heat of vaporization of toluene.

$$L = 230.33\left(1 - \frac{T}{591.7}\right)^{0.38} = 230.33\left(1 - \frac{484.75}{591.7}\right)^{0.38} = 120.24\ Btu/lb$$

Step 6. Calculate relief load.

$$Relief\ load = \frac{Fire\ heat\ input}{Latent\ heat\ at\ relief\ condition} = \frac{2.081 \times 10^{6}\ Btu/h}{120.24\,Btu/lb} = 1.731 \times 10^{4}\,lb/h$$

Step 7. Check for critical flow

$$P_{critical} = P_{relief}\left(\frac{2}{n+1}\right)^{\frac{n}{n-1}} = 135.7 \times \left(\frac{2}{1.0839+1}\right)^{\frac{1.0839}{1.0839-1}} = 79.8\ psia$$

Since this is greater than the back pressure, flow is critical.

Step 8. Calculate factor C

$$C = 520\sqrt{n\left[\frac{2}{n+1}\right]^{\frac{n+1}{n-1}}} = 520\sqrt{1.0839\left[\frac{2}{1.0839+1}\right]^{\frac{1.0839+1}{1.0839-1}}} = 324.98$$

where n = ratio of specific heats = 1.0839, given by the problem.

Step 8. Calculate relief area.

$$A = \frac{W}{CKP_1K_b}\sqrt{\frac{TZ}{M}} = \frac{1.731 \times 10^{4}}{324.98 \times 0.975 \times 135.7 \times 1}\sqrt{\frac{872.6 \times 0.831}{92}}$$

$= 1.13\ in^2$. Select valve: 2J3

16-3

Step 1. Calculate relief mass flux, G_T

The first equation will be used to calculate relief mass flux. Two parameters are unknown. The first one ψ will be assumed 1 since we do not know the diameter and the length of relief piping. The second T_s will be determined from the given Antoine equation.

Set pressure = 114.7 psia = 5930.1 mmHgA

$$T_s = 47.181 + \frac{3242.4}{16.266 - \ln(5930.1)} = 475\,K$$

$$G_T = 0.9 x 1 x \frac{2.826 x 10^5}{0.03734} \sqrt{\frac{1}{2509.6 \times 475} \frac{N.m}{kg} \frac{kg}{m^3}} \sqrt{\frac{kg.m}{s^2.N} \frac{kg.K}{N.m} \frac{1}{K}}$$

$= 6238.7\ kg/m^2.s$ [note : J = N.m]

Step 2. Calculate relief area, A

This requires manipulation of the second equation. First, three parameters are to be evaluated: m_0, V, and Q.

Shell volume = $\pi(8)^2(15)/4$ = 753.982 cuft
Bottom volume = 501.3 gallons = 67.019 cuft
Initial volume = 753.982 + 67.019 = 821.001 cuft = 23.234 m^3
Gross volume of vessel, V = 753.982 + 67.019x2 =888.02 cuft = 25.142 m^3
Initial mass = m_0 = 23.234x655.2067 = 15223.072 kg
Constant heat input = Q = 2.081×10^6 Btu/h = 609,848 J/s

Substituting the known values of the parameters in the second equation:

$$A = \frac{Q m_0 V_{fg}}{G_T V (\Delta H_v)} = \frac{609848 \times 15223.072 \times 0.03734}{6238.7 \times 25.142 \times 2.826 \times 10^5}$$

$= 0.00782\ m^2 = 12.12\ in^2$

Select relief valve : 6R8

MORNING SAMPLE EXAMINATION

Instructions for Morning Session

1. You have four hours to work on the morning session. Do not write in this handbook. Write your answers only on the score sheets provided.

2. Work only four questions. Only four questions will be scored.

3. Work rapidly and use your time effectively. If you do not know the correct answer, skip it and return to it later.

4. Some problems are presented in both metric and English units. Solve either problem.

Special NOTE:

This exam is only half of the length of the standard exam with 5 morning problems. You are to solve two complete problems in two hours, or choose to solve four complete problems in four hours. Select the problems carefully, as only four problems will be scored. You will be provided with score sheets for your answers. Be sure to sign each sheet.

1 : An oil-fired steam boiler produces saturated steam at 200 psig. Oil firing rate is 5811 lb/h. Boiler feed water is at 190 ^{0} F . The flue gases leave the furnace at 500 ^{0} F and the radiation and convection losses from the furnace can be taken to be 15% of the total heat input. Calculate the amount of steam generated per hour and overall thermal efficiency of the station. Additional data are as follows:

Composition of oil fired: C = 86.47% H = 11.65 % S = 1.35% O = 0.27 %
N = 0.26 %
HHV = 19178 Btu/lb
Air supplied to the burner: temperature 70 ^{0} F Pressure = 755 mm Hg
% excess air = 20

2: A single effect evaporator is to concentrate 30000 lb/h of NaOH solution from 10 to 50 % concentration. The evaporator is to use 15 psig steam. The feed is at 100 ^{0}F. It operates under a vacuum of 26 in Hg. If overall coefficient is 425 Btu/h-ft^2 -^{0}F. Calculate

(1) heat transfer surface
(2) steam economy and
(3) water requirement of a countercurrent barometric condenser. if water is available at 86 ^{0}F and the outlet water temperature is not to exceed 110 ^{0}F

3: 200 kg of a mixture of 35 % chloroform and 65% acetic acid are thoroughly mixed with 100 kg of water. (a) Determine the compositions and amounts of the extract and raffinate phases. Conjugate phase data and the distribution diagram are given in Figure 3.a and b. If the raffinate in the first stage is treated with half its weight of water, calculate the cumulative recovery of acetic acid in the two stages.

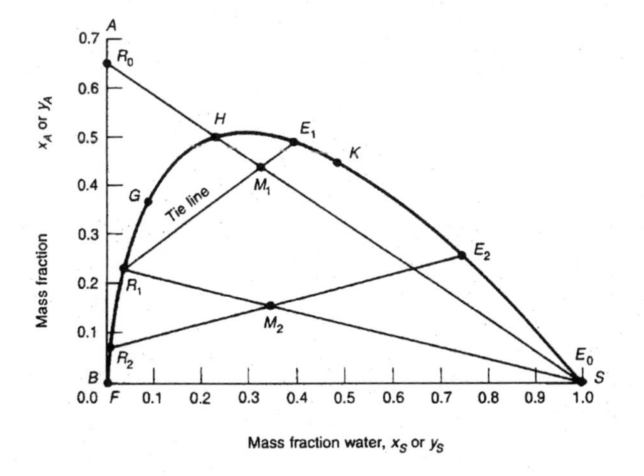

Figure 3 a

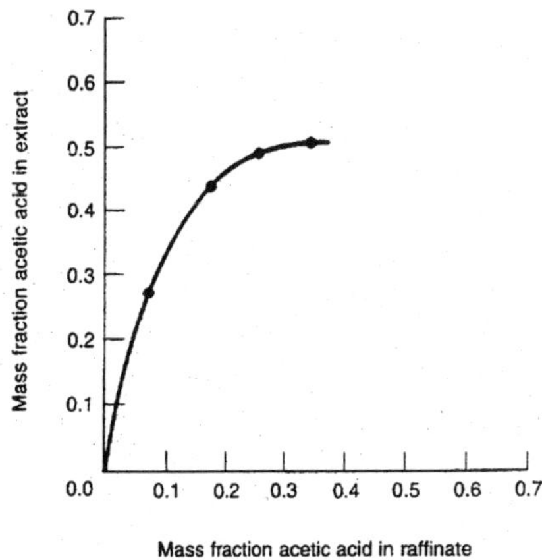

Figure 3 b

4: A wet solid is dried from 40 to 15 % moisture in 5 hours under constant drying conditions. The critical moisture is 20 % and the equilibrium moisture is 5 %. Calculate

(a) The additional time that it would take to dry the material to 10 % moisture

(b) If the solid , a 5 cm thick slab is to be dried from both sides and has a density of 1920 kg/m^3 , what is the drying rate at moisture content of 10 %.

5: (a) Calculate the entropy change of SO_2 when it is cooled from 538 to -101 ^0C at a constant pressure of 1 atm. Express the result as Btu/(lb) (^0R). The molar heat capacity of SO_2 is given by

$$C_P = 6.517 + 1.38 \times 10^{-2} T + 0.9103 \times 10^{-5} T^2 + 2.057 \times 10^{-9} T^3$$

(b) Calculate the absolute entropy of solid SO_2 at its melting point at a pressure of 1 atm. Express the result as Btu/(lb) (^0R).

Data on SO_2 at 1 atm. pressure:

Boiling point = - 5.0 ^0C Latent heat of vaporization = 5960 g-cal/g-mole

Melting point = -75.5 ^0C Latent heat of fusion = 1769 g-cal/g-mole

Specific heat of liquid SO_2 = 0.310 g-cal/(gram) (K)

Specific heat of solid SO_2 = 0.229 g-cal/ (gram) (K)

Solutions

1.

Basis: 100 lb of oil fired

Calculation of flue gases

component	lb	lb-moles	O_2 required
carbon	86.47	7.21	7.21
hydrogen	11.65	5.83	2.915
sulfur	1.35	0.042	0.042
nitrogen	0.26	0.01	-
oxygen	0.27	0.01	- 0.01

Total oxygen required = 10.157 lb-mol

Excess oxygen = 0.2 x 10.157 = 2.031 lb-mols

Dry air used = (10.157 + 2.031) /0.21 = 58.04 lb-mol.

Nitrogen in air = 0.79 x 58.04 = 45.85 lb-mol

Flue gases :

Basis : 100 lb coal

	lb-mol	Av,. C_P
CO_2	7.21	8.894
H_2O	5.83	8.026
SO_2	0.04	9.54
N_2	45.85	6.961
O_2	2.03	7.09
Total	60.96	

$$\text{Average } C_P = \frac{7.21\times8.894+5.83\times8.026+0.04\times9.54+45.85\times6.96+2.03\times7.019}{60.96} = \text{Btu/lb-mol } ^0\text{F}$$

Calculate net heating value of fuel

HHV of oil = 19718 Btu/lb

Hydrogen in 1 lb of oil = 0.1165 lb = 0.05825 lb-mol

Water formed = 0.05825 lb-mol = 1.0485 lb/lb of oil

Latent heat of vaporization of water at 77 ^{0}F = 1050.1 Btu/lb

Therefore , LHV = 19718 - 1.0485 x 1050.1 = 18077 Btu /lb oil fired.

Heat balance on the furnace

Heat of combustion of 100 lb oil = 100 x 18077 = 1807700 btu/100 lb oil

Heat lost with gases = 60.96 x (7.3) (500 - 70) = 191353 Btu/100 lb oil

Other heat losses = 0.15 x 1807700 = 271155 Btu

Total heat losses = 191353 + 271155 = 462508 Btu
Heat utilized in steam production = 1807700 - 462508 = 1345192 Btu
Overall thermal efficiency = (1345192/1807700) x 100 = 74.4 %

Oil firing rate = 5811 lb/h

Therefore, heat released = (5811/100) x 1807700 = 105.05 x10^6 ^{0}F Btu/h
Heat used in steam generation = 0.744 x 105.05 x 10^6 = 78.16 x 10^6 Btu/h
From steam tables,
Enthalpy of water at 190 ^{0}F = 158.04 Btu/lb
Enthalpy of steam at 200 psig = 214.7 psia = 1200.1 Btu/lb
Heat required to generate 1 lb of steam = 1200.1 - 158.04 = 1042.06 Btu/lb
Boiler capacity = (78.16 x 10^6/1042.06) = 75000 lb/h of steam at 200 psig

2:

Material Balance

F = 30000 lb/h x_F = 0.1 t_F = 100 ^{0}F x_L = 0.5

$F\,x_F = L\,x_L + Vy_S$ $y_S = 0$ in vapor

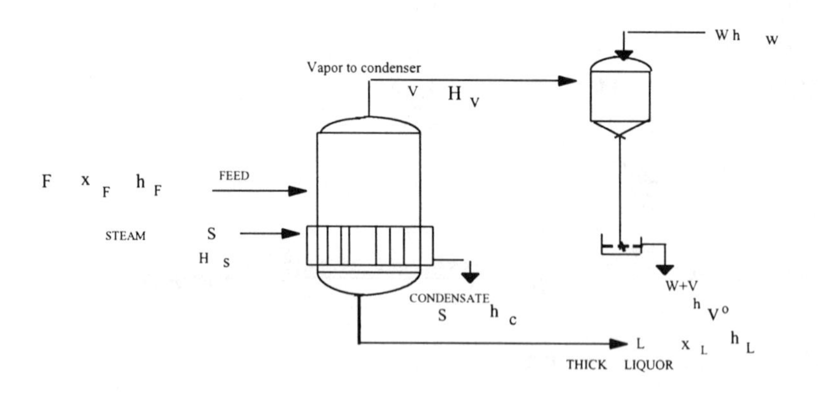

30000 (0.1) = L(0.5) therefore L = 6000 lb/h NaOH in solution = 3000 lb/h
Water to be evaporated = 30000 - 6000 = 24000 lb/h

Enthalpy Balance

Steam condition and properties: Steam pressure = 15+ 14.7 = 29.7 psia
t_S = 250 ^{0}F H_S = 1164 Btu/lb = 945.5 Btu/lb

The boiling point of water under 26 in Hg vacuum = 4 in Hg absolute pressure
= (4/29.92) x 14.7 = 1.96 psia

From steam tables, t_V = 125 ^{0}F H_V for saturated steam = 1116 Btu/lb

Enthalpies for NaOH solutions are to be obtained from enthalpy composition diagram and the boiling point of solution to be obtained from a Dühring's plot for NaOH. These are available elsewhere. From Dühring's plot the boiling point of 50 % solution is 198 ^{0}F. From enthalpy composition diagram, h_F = 60 Btu/lb and h_L = 222 Btu/lb.

Energy balance gives $S (H_S - h_C) = VH_V + Lh_L - F\ h_F$

H_V = 1116 + 0.46(198 - 125) = 1149 Btu/lb at 198 ^{0}F includes superheat. h_C = 218.5 Btu/lb

Substitution gives S(1164 - 218.5) = 24000(1149) + 6000(222) - 30000 (60)

From which S = 28670 lb/h

Heating surface:

U = 425 Btu/h-ft^2- ^{0}F = t_S - t_B = 250 - 198 = 52 ^{0}F

Q = 28670(1164-218.5) = 27107485 Btu/h

Therefore A = 27107485/(425 x 52) = 1226 ft^2

Steam economy = 24000/28670 = 0.837 lb/lb of steam

Condenser water $V\ H_V + Wh_{wi} = (W + V)\ h_{WO}$

24000 x 1149 + W x 54.03 = (W + 24000) x 87.97

Solving for W, W = 750286 lb/h = 1500.6 gpm.

3:

We shall use the diagram given in figure 3a to make the graphical construction.

(a) Locate feed point R_0 (A=0.65, B= 0.35 and S = 0) on the given diagram.
Also locate solvent water as point S where mass fraction of water = 1.
Total mixture after addition of water = 200 + 100 = 300 kg
Therefore the coordinates of the addition point are:

$\left(x_A\right)$ (130/300) = 0.433 and $\left(x_S\right)_{M_1}$ = 100/300) = 0.333

(all compositions in mass fractions)

Locate this point on $\overline{R_0S}$ as M1. With the help of the distribution diagram, locate the tie line that passes through M_1. The ends of this tie line give the compositions of E_1 and R_1.

Read from the diagram the compositions of extract and raffinate as follows

	Extract	kg	Raffinate	kg
	0.480	117.1	0.230	12.9
	0.395	96.4	0.030	1.7
	0.125	30.5	0.740	41.4
Total	1.00	244.0	1.00	56.00

$$R_1 + E_1 = 300 \text{ kg}$$

$$0.48\, E_1 + 0.23\, R_1 = 130 \text{ by solute balance}$$

Solving $R_1 = 56$ kg $E_1 = 244$ kg

Calculate the amounts of individual components in raffinate and first extract. These are given in the table of compositions above.

(Note: Ratio of R_1 to E_1 can be obtained from graph as $E_1M_1/ R_1 M_1$ and this can be used to determine the amounts of E_1 and R_1)

(b) New amount of mixture after addition of fresh water

	kg	mass fraction
Solute A	= 12.9	0.154
Water	= 1.7 + 28 = 29.7	0.354
$CHCl_3$	= 41.4	0.492
Total	84	

This point is located on M_2 and a tie line is drawn through M_2 to give E_2 ,and R_2
From the diagram, the compositions of E_2 and R_2 are

	E_2	R_2
x_A	0.250	0.065
x_S	0.74	0.011
x_B	0.01	0.924

$$E_2 + R_2 = 84$$

$$0.25\, E_2 + 0.065\, R_2 = 12.9$$

Solving $E_2 = 40.2$ and $R_2 = 43.8$ kg
Solute recovered in two extracts = 244(0.48) + 43.8(0.065) = 120 kg
Recovery = (120/130) x 100 = 92.3 %

4:

(a) $\theta_C + \theta_f = \frac{L_S}{AN_c}\left(X_1 - X_c\right) + \frac{L_S}{AN_C}(X_c - X*)\ln\frac{X_c - X*}{X_2 - X*} = 5$ hr

$X_1 = \frac{0.4}{1-0.4} = 0.67$ $X_c = \frac{0.2}{1-0.2} = 0.25$ kg H_2O/kg dry solid.

$X_2 = \frac{0.15}{1-0.15} = 0.1765$ $X_3 = \frac{0.1}{0.9} = 0.1111$ $X* = \frac{0.05}{0.95} = 0.0526$

$$\frac{L_S}{AN_C}\left[\left(X_1 - X_c\right) + (X_c - X*)\ln\frac{(X_c - X*)}{(X_2 - X*)}\right] = 5$$

Substituting appropriate values of the variables,

$$\frac{L_S}{AN_c}\left[(0.67 - 0.25) + (0.25 - 0.0526)\ln\frac{0.25 - 0.0526}{0.1765 - 0.0526}\right] = 5$$

Simplifying gives

Therefore $\frac{L_S}{AN_C} = (5/0.511924) = 9.77$

$$\theta_{f1} = 9.77[X_c - X*]\ln\frac{0.25 - 0.0526}{0.1765 - 0.0526} = 0.9$$

$$\theta_{f2} = 9.77(0.25 - 0.0526)\ln\frac{0.25 - 0.0526}{0.1111 - .0.0526} = 2.35 \text{ h}$$

$$\theta_C = 9.77(0.67 - 0.25) = 4.1 \text{ h}$$

The additional time taken to dry from 15 to 10 % moisture is 2.35 - 0.9 = 1.45 h

(b) Volume = $5\text{x}10^{-2}$m (1 m^2) = $5\text{x}10^{-2}$ m^3
L_S = $5\text{x}10^{-2}$ (1920) = 96 kg
Slab to be dried from both sides $L_S/A = 96/2 = 48$ kg/m^2

$\frac{L_S}{AN_c} = 9.77$ $N_c = \left(L_S/A\right)/9.77 = 48/9.77 = 4.913$ kg/h.m^2

N - drying rate at 10 % moisture content = $\frac{N_C(X - X*)}{X_c - X*} = \frac{4.913(0.111 - 0.0526)}{0.25 - 0.0526} = 1.45$ kg/hm^2

5:

SO_2 (g) $\rightarrow$ SO_2(l) $\rightarrow$ SO_2 (s)
$T_1 = 537 + 273 = 810$ K $T_C = -5 + 273 = 268$ K $T_f = -75.5 + 273 = 197.5$ K
Basis 1 g-mol of SO_2

Gas Cooling

$$S = \int_{810}^{268} \frac{C_P dT}{T} = \int_{810}^{268} \frac{6.517 + 1.38x10^{-2}T - 0.9103x10^{-5}T^2 + 2.057x10^{-9}T^3}{T}$$

$$= 6.517 \ln \frac{T_2}{T_1} + 1.38x10^{-2}(T_2 - T_1) - \frac{0.9103x10^{-5}}{2}(T_2^2 - T_1^2) + \frac{2.057x10^{-9}}{3}(T_2^3 - T$$

$$= 6.517 \ln \frac{268}{810} + 1.38x10^{-2}(268 - 810) - 0.45515x10^{-5}(2682 - 810^2) + 0.6856x10^{-9}(268^3 - 810^3)$$

$$= -7.2081 - 7.4796 + 2.6593 - 0.3512$$
$$= -12.38 \text{ g-cal/(g-mol)(K)}$$

Gas condensation

$$\Delta S = \frac{-Q}{T} = \frac{-5960}{268} = -22.2389 \text{ g-cal/(g-mol)(K)}$$

Liquid Cooling

$$\Delta S = \int_{268}^{197.5} \frac{C_P dT}{T} = (64)(0.310) \int_{268}^{197.5} \frac{dT}{T} = 64x0.31 \ln \frac{197.5}{268} = -6.056 \text{ g-cal/(g-mol)(K)}$$

Phase change from liquid to solid

$$\Delta S = \frac{-1769}{197.5} = -8.957 \text{ g-cal/(g-mol)(K)}$$

Cooling of solid

$$\Delta S = \int_{197.5}^{173} \frac{0.229x64dT}{T} = 14.656 \int_{197.5}^{173} \frac{dT}{T} = \ln \frac{173}{197.5} = -1.941 \text{ g-cal/(g-mol)(K)}$$

Total change in entropy = -12.38 - 22.34 - 6.056 - 8.957 - 1.9411
= -29.778 g-cal/(g-mol)(K)

$$= -29.778 \left[\frac{\frac{gcal}{252\, gcal/Btu}}{\frac{gmol \times 64\, g}{454\, g/lb} \times K \times 1.8\ {}^0R/K} \right] = 0.4653 \text{ Btu/(lb)(}^0\text{R)}$$

(b) Since entropy is a state property, we can find ΔS that takes place in cooling from 25 ^{0}C to the melting point and knowing the absolute entropy at 25 ^{0}C, calculate the absolute entropy at the melting point.

Entropy change in cooling the gas from 25 ^{0}C to condensation point

$$\Delta S = 6.517 \ln \frac{268}{298} + 1.38x10^{-2}(268 - 298) - (0.9103/2)x10^{-5}(268^2 - 298^2) + (2.057/3)x10^{-9}(268^3 - 298^3)$$

$= -1.0332$ gcal/(gmol) (K)

Then total entropy change from 25 ^{0}C to melting point is given by

$\Delta S = -1.0332 - 22.34 - 6.056 - 8.957 = -38.2851$ gcal/(gmol)(K)

Then absolute entropy at the mp $= 59.4 - 38.2851 = 20.8549$ gcal/(gmol)(K)
$= 0.3258$ Btu/(lb) (^{0}R)

AFTERNOON SAMPLE EXAMINATION

Instructions for Afternoon Session

1. You have four hours to work on the afternoon session. Do not write in this handbook.

2. Answer only four questions for a total of forty answers. There is no penalty for guessing.

3. Work rapidly and use your time effectively. If you do not know the correct answer, skip it and return to it later.

4. Some problems are presented in both metric and English units. Solve either problem.

5. Mark your answer sheet carefully. Fill in the answer space completely. No marks on the workbook will be evaluated. Multiple answers receive no credit. If you make a mistake, erase completely.

Work 4 afternoon problems in four hours.

Special NOTE:

This exam is only half of the length of the standard exam with 10 multiple choice problems, each with 10 questions. You are to solve two complete problems in two hours, or choose to solve four complete problems in four hours. Select the problems carefully, as only four problems will be scored.

P.E. Chemical Engineering Exam
Afternoon Session

Ⓐ Ⓑ Ⓒ Fill in the circle that matches your exam booklet.

1-1 Ⓐ Ⓑ Ⓒ Ⓓ	2-1 Ⓐ Ⓑ Ⓒ Ⓓ	3-1 Ⓐ Ⓑ Ⓒ Ⓓ	4-1 Ⓐ Ⓑ Ⓒ Ⓓ	5-1 Ⓐ Ⓑ Ⓒ Ⓓ
1-2 Ⓐ Ⓑ Ⓒ Ⓓ	2-2 Ⓐ Ⓑ Ⓒ Ⓓ	3-2 Ⓐ Ⓑ Ⓒ Ⓓ	4-2 Ⓐ Ⓑ Ⓒ Ⓓ	5-2 Ⓐ Ⓑ Ⓒ Ⓓ
1-3 Ⓐ Ⓑ Ⓒ Ⓓ	2-3 Ⓐ Ⓑ Ⓒ Ⓓ	3-3 Ⓐ Ⓑ Ⓒ Ⓓ	4-3 Ⓐ Ⓑ Ⓒ Ⓓ	5-3 Ⓐ Ⓑ Ⓒ Ⓓ
1-4 Ⓐ Ⓑ Ⓒ Ⓓ	2-4 Ⓐ Ⓑ Ⓒ Ⓓ	3-4 Ⓐ Ⓑ Ⓒ Ⓓ	4-4 Ⓐ Ⓑ Ⓒ Ⓓ	5-4 Ⓐ Ⓑ Ⓒ Ⓓ
1-5 Ⓐ Ⓑ Ⓒ Ⓓ	2-5 Ⓐ Ⓑ Ⓒ Ⓓ	3-5 Ⓐ Ⓑ Ⓒ Ⓓ	4-5 Ⓐ Ⓑ Ⓒ Ⓓ	5-5 Ⓐ Ⓑ Ⓒ Ⓓ
1-6 Ⓐ Ⓑ Ⓒ Ⓓ	2-6 Ⓐ Ⓑ Ⓒ Ⓓ	3-6 Ⓐ Ⓑ Ⓒ Ⓓ	4-6 Ⓐ Ⓑ Ⓒ Ⓓ	5-6 Ⓐ Ⓑ Ⓒ Ⓓ
1-7 Ⓐ Ⓑ Ⓒ Ⓓ	2-7 Ⓐ Ⓑ Ⓒ Ⓓ	3-7 Ⓐ Ⓑ Ⓒ Ⓓ	4-7 Ⓐ Ⓑ Ⓒ Ⓓ	5-7 Ⓐ Ⓑ Ⓒ Ⓓ
1-8 Ⓐ Ⓑ Ⓒ Ⓓ	2-8 Ⓐ Ⓑ Ⓒ Ⓓ	3-8 Ⓐ Ⓑ Ⓒ Ⓓ	4-8 Ⓐ Ⓑ Ⓒ Ⓓ	5-8 Ⓐ Ⓑ Ⓒ Ⓓ
1-9 Ⓐ Ⓑ Ⓒ Ⓓ	2-9 Ⓐ Ⓑ Ⓒ Ⓓ	3-9 Ⓐ Ⓑ Ⓒ Ⓓ	4-9 Ⓐ Ⓑ Ⓒ Ⓓ	5-9 Ⓐ Ⓑ Ⓒ Ⓓ
1-10 Ⓐ Ⓑ Ⓒ Ⓓ	2-10 Ⓐ Ⓑ Ⓒ Ⓓ	3-10 Ⓐ Ⓑ Ⓒ Ⓓ	4-10 Ⓐ Ⓑ Ⓒ Ⓓ	5-10 Ⓐ Ⓑ Ⓒ Ⓓ

1: Barium Sulfide is the starting material for the production of various Barium chemicals. It is industrially produced from barites by reduction of $BaSO_4$, its main component after beneficiation of the ore. The accompanying block flow diagram Matbal 1 illustrates the process scheme followed by a plant manufacturing $BaCO_3$, and $BaCl_2$. The average composition of the black ash exiting the rotary kiln is also given in the diagram. Other data are as follows:

Normal temperature at site = 70 °F.

Atmospheric pressure = 755 mm Hg

Figure 1 : Production of Barium Chemicals from Barites.

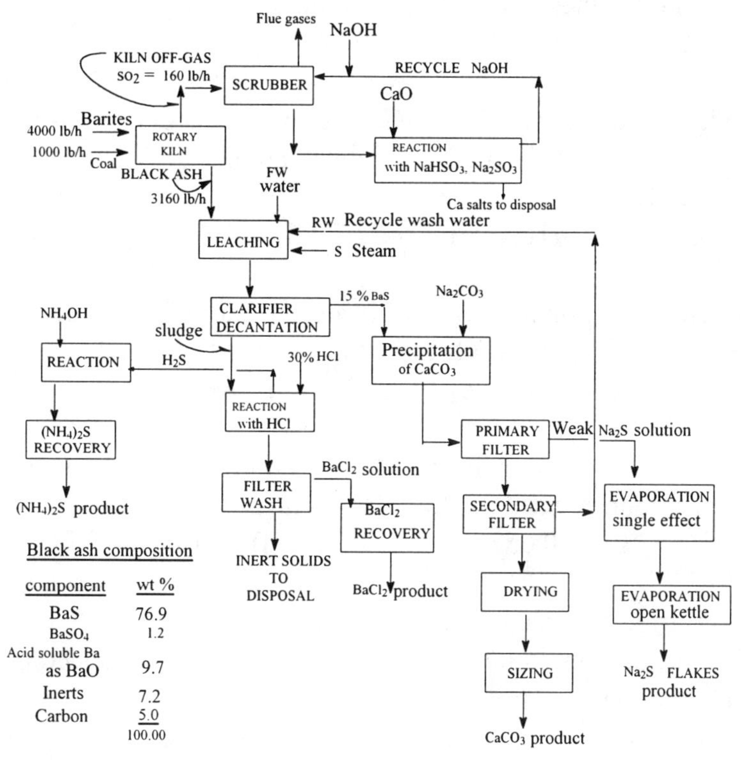

Black ash composition

component	wt %
BaS	76.9
$BaSO_4$	1.2
Acid soluble Ba as BaO	9.7
Inerts	7.2
Carbon	5.0
	100.00

Molecular weights :

Barium sulfate	233.4	Barium sulfide	169.4	Barium oxide	153.4	Barium carbonate	197.4
Sodium carbonate	106	Sodium hydroxide	40	Ammonium sulfide	68	Sodium sulfate	142
Sodium sulfite	126	Sodium bisulfite	104	Hydrochloric acid	36.5	Barium chloride	208.4
Sulfur dioxide	64	Sodium sulfide	78				

Equations :

Barium sulfate reduction: $BaSO_4 + 4CO = BaS + 4CO_2$

Reaction of BaS with HCl: $BaS + 2\,HCl = BaCl_2 + H_2S$

Precipitation of Barium carbonate: $BaS + Na_2CO_3 = BaCO_3 + Na_2S$

Reaction of H_2S with NH_4OH : $H_2S + 2\,NH_4OH = (NH_4)_2S + 2H_2O$

Sodium bisulfite formation in scrubber: $SO_2 + NaOH = NaHSO_3$

Sodium sulfate formation in scrubber: $SO_2 + 2\,NaOH = Na_2SO_3 + H_2O$

$Na_2SO_3 + 1/2\,O_2 = Na_2SO_4$

1.1 If Barium occurs in the beneficiated ore as $BaSO_4$ only, the percentage by weight of $BaSO_4$ in the beneficiated ore is most nearly:

a) 95.4 b) 96.3 c) 83.7 d) 90

1.2 Pounds of carbon used in the reduction of $BaSO_4$ is nearly:

a) 788.5 b) 800 c) 396 d) 792.5

1.3 The blackash is treated with hot water and clarified in a Dorr thickener. If the clean overflow solution from the thickener contains 15 % BaS and 65 % of BaS in the blackash is recovered in the overflow, pounds of solution retained per lb of water-insoluble solids in the thick sludge is closest to:

a) 7.8 b) 11.3 c) 8.6 d) 9.14

1.4 If the thick sludge is treated with 30% HCl acid in 5 % excess and if the conversion of all HCl soluble Ba compounds is 100 % complete, pounds of $BaCl_2$ that can be obtained per hour is nearly:

a) 1045.6 b) 416.4 c) 1462.7 d) 1080.2

1.5 If the clear overflow solution is treated with sodium carbonate and if the reaction is 95% complete, the amount (lb) of $BaCO_3$ that will precipitate is nearly :

a) 2830 b) 1500 c) 1748.67 d) 1840.6

1.6 If in the item 5 above, the sodium carbonate is used in 5 % excess over theoretical, the pounds of sodium carbonate used per lb of Bas is nearly:

a) 0.626 b) 0.657 c) 1.22 d) 1.0

1.7 If the hydrogen sulfide evolved in the reaction between BaS and HCl is absorbed in a solution of NH_4OH , the amount in tons of $(NH_4)_2S$ that can be produced per year with on-stream factor of 8000 hours/year is nearly:

a) 1365.6 b) 682.8 c) 1229 d) 1200

1.8 In order to precipitate $BaCO_3$, 30 % Na_2CO_3 solution is added in 5 % excess to BaS solution. The $BaCO_3$ precipitated is filtered and washed with 1000 lb of water on a continuous basis. If the reaction is 95 % complete, the percentage of Na_2S by weight in the resultant filtrate is nearly:

a) 5.61 b) 5.72 c) 5.22 d) 6.5

1.9 The dilute Na_2S solution in item 8 is generally evaporated in two stages. In the first stage, the solution is first concentrated to 30 % by evaporation in a single evaporator equipped with a steam-heated coil. The 30 % solution is then concentrated to 62 % in an open direct fired kettle. The preferred material of construction for the steam coil in the single stage evaporator is:

a) Carbon steel b) 304 SS c) Monel d) 316 SS

1.10 The flue gases from the barite reduction rotary kiln usually contain sulfur dioxide due to the reaction of BaS with O_2 and the combustion of sulfur in the fuel. If the rate of sulfur dioxide generation is 160 lb/h, and 10 % of the SO_2 is oxidized to sulfate in the scrubber, the tons of NaOH/day, that will be consumed in the scrubber will be closest to:

a) 2.1 b) 1.32 c) 1.7 d) 0.9

2: Oil is to be extracted from livers by continuous countercurrent multistage extraction with ether. The locus of the underflow compositions is given in figure 2. The livers contain 0.3 mass fraction of oil. The fresh solids charge is 1000 lb/h and 95 % of oil is to be recovered. Assume fresh solvent to be pure. The efficiency is 75 %. On the basis of these data, answer the following questions.

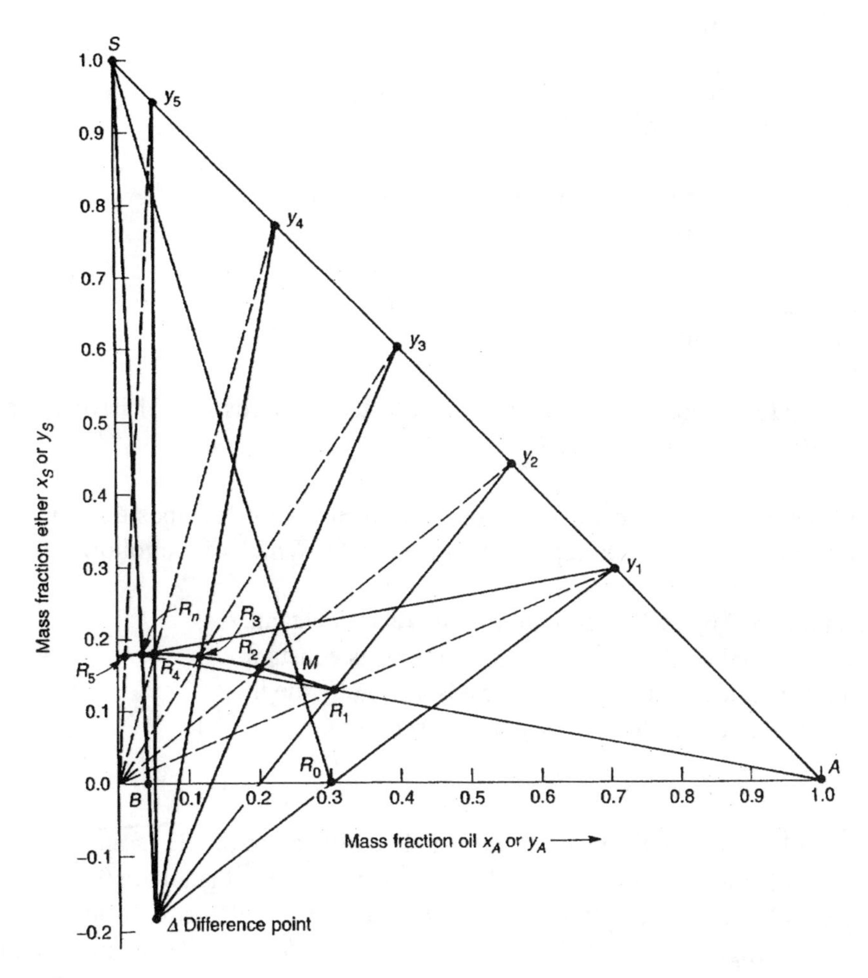

Figure 2.

2.1 Mass fraction of oil in discharged solids excluding the solvent is very nearly
a) 0.3 b) 0.033 c) 0.19 d) 0.0411

2.2 The quantity of the solvent in the discharged solids in lb/h is very nearly
a) 180 b) 125.3 c) 159 d) 131

2.3 The minimum solvent-to-solids ratio that can be used to carry out the separation is
a) 0.17 b) 0.25 c) 0.26 d) 0.14

2.4 If the solvent amount actually used is 1.6 times the minimum required , ether to be fed in lb/h is
a) 190 b) 200 c) 272 d) 250

2.5 The coordinates of the addition point are
a) $(x_A)_M = 0.236$, $(x_S)_M = 0.214$, $(x_S)_M = 0.55$
b) $(x_A)_M = 0.2138$, $(x_S)_M = 0.5504$, $(x_S)_M = 0.2358$
c) $(x_A)_M = 0.5504$, $(x_S)_M = 0.2358$, $(x_S)_M = 0.2138$
d) $(x_A)_M = 0.18$, $(x_S)_M = 0.21$, $(x_S)_M = 0.61$

2.6 Mass fraction of ether in the solution leaving the first stage is
a) 0.09 b) 0.3 c) 0.25 d) 0.295

2.7 The number of theoretical stages required is very nearly
a) 5 b) 4 c) 4.4 d) 6

2.8 The number of actual stages to be recommended would most likely be
a) 6 b) 5.9 c) 5 d) 7

2.9 Net flow of the solute (component A) in lb/h from nth stage towards stage 1 is very nearly
a) +270 b) -30 c) +30 d) -270

2.10 The data on a leaching system show that the locus of the underflow compositions is a straight line and passes through the vertex representing the solute. Which of the following statements is true?
a) The solvent retained by the inerts in lb/lb of inerts is constant.
b) The solution retained by the inerts in lb/lb of inerts is constant.
c) The system is nonideal and there is solute adsorption by the inerts
d) The tie lines will not pass through origin representing inerts. (I = 1).

3: A mixture of ethanol vapor and nitrogen has a dry-bulb temperature of 122 °F and wet-bulb temperature of 68 °F. at std atm. pressure. Specific heats of nitrogen and ethanol are 0.25 and 0.38 Btu/(lb.°F) respectively. Latent heat of evaporation for alcohol is 16910 Btu/lb-mol. Diffusivity of alcohol in nitrogen = 0.44 ft²/ h. Vapor pressures of alcohol = 43.6 mmHg at 20 °C and 229 mmHg at 122 °F.

Viscosity of air = 0.018 cP, $\frac{h_G}{k_y}$ = 0.379 Btu/(lb. °F).
Given these data, select the correct answer for each of the following questions:

3.1 Wet-bulb absolute saturation in units of lb/lb dry air is
a) 0.09 b) 0.1 c) 0.12 d) 0.05

3.2 The value of the Schmidt number is very nearly
a) 1.0 b) 1.36 c) 1.32 d) 0.565

3.3 The dry-bulb absolute saturation lb alcohol/ lb nitrogen is very nearly
a) 0.0443 b) 0.100 c) 0.5 d) 0.056

3.4 Saturated absolute saturation in lb alcohol/lb nitrogen is very nearly
a) 0.80 b) 0.709 c) 0.75 d) 0.045

3.5 The percentage saturation for the mixture is very nearly
a) 6.25 b) 7.09 c) 10 d) 8.9

3.6 The percent relative saturation of the mixture is very nearly
a) 4.9 b) 8.7 c) 5.5 d) 6.4

3.7 Humid volume of the mixture in ft³/lb is very nearly
a) 14.13 b) 13.17 c) 12.82 d) 17.3

3.8 The humid heat of the mixture in Btu/lb Dry air is very nearly
a) 0.294 b) 0.267 c) 0.25 d) 0.38

3.9 The value of Lewis number for the mixture is very nearly
a) 2.1 b) 1.315 c) 1.92 d) 2.6

3.10 Heat in Btu/h needed to heat 1000 ft³/min of the mixture from 122 to 220 °F is very nearly
a) 111108 b) 10000 c) 22000 d) 11100

4: A refrigeration unit used by a plant supplies 506.2 gpm of glycol brine (20 % Ethylene glycol in water) at 30 °F and receives the hot brine from the users at 40 °F. Cooling water is available at 85 °F. The equipment design allows 10 °F approach in both the condenser and the evaporator. The plant currently uses R-12 as refrigerant but wants to switch to an environmentally safe HFC-134a. Operating conditions and some data are shown in Figure Thermo-1 and the relevant thermodynamic properties of HFC -134a are given in Table Thermo 1. The glycol brine has a specific gravity of 1.02. Its specific heat is 0.93 Btu/lb °F .

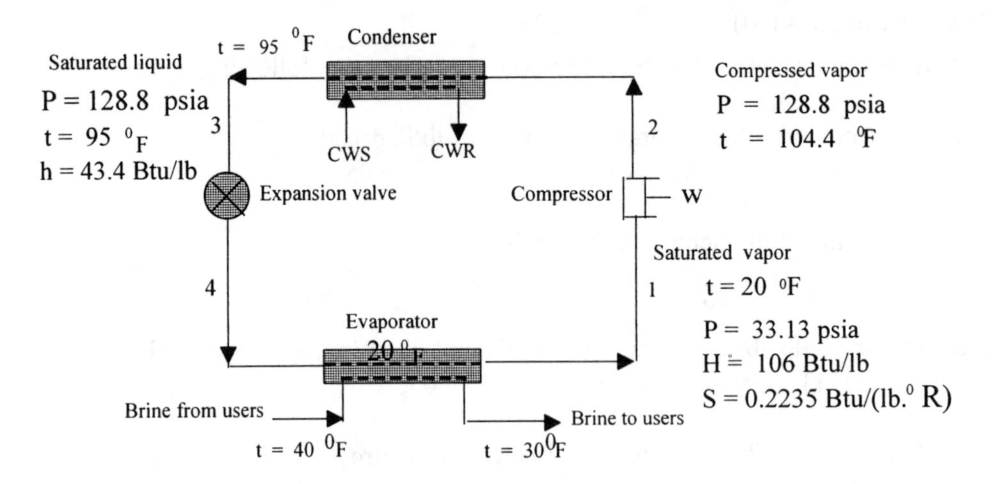

Table 4: Thermodynamic Property Data

Properties of Saturated Refrigerant HFC-134a (Eng. Units) Temperature Table

, cm³/g; U, Btu/lb; H, Btu/lb; S, Btu/(lb.°F)

		Specific volume		Internal energy		Enthalpy			Entropy	
Temp. °F t	Press. psia P	Sat. Liquid v_l	Sat. Vapor v_g	Sat. Liquid U_l	Sat. Vapor U_g	Sat. Liquid h_l	Evap. λ	Sat. Vapor H_g	Sat. Liquid S_l	Sat. Vapor S_g
20	33.13	0.0122	1.409			18.4	87.6	106.0	0.0408	0.2235
95	128.8	0.0137	0.369			43.4	72.7	115.7	0.0880	0.2193

Properties of superheated Refrigerant HFC-134a (Eng Units)

V, cm³/g; U, Btu/lb; H, Btu/lb; S, Btu/(lb.°F)

	Pressure 120 psia (t = 90.5 0F) Temp.				Pressure 130 psia (t = 95.6 °F)			
°F		U	H	S		U	H	S
100	0.041044		117.7	0.2238	0.37136		117.0	0.2212
110	0.42402		120.2	0.2282	0.38453		119.5	0.2257

With the data given in the tables, answer the following questions by selecting the correct answer.

4.1 The capacity of the refrigeration unit in tons of refrigeration is
a) 316 b) 325 c) 350 d) 300

4.2 The heat removal in the evaporator in Btu/lb of refrigerant is
a) 40 b) 62.6 c) 50 d) 60

4.3 The mass flow rate of the refrigerant in lb/min is
a) 1000 b) 958.5 c) 685 d) 82

4.4 The work of compression in Btu/lb done is
a) 36.3 b) 62.6 c) 12.06 d) 74.7

4.5 The coefficient of performance of the unit is
a) 5.19 b) 6.47 c) 6.3 d) 5.5

4.6 The Carnot coefficient of performance of the reversible refrigeration cycle is
a) 4.9 b) 6.4 c) 5.19 d) 5.5

4.7 Vapor mass fraction of the refrigerant after valve expansion is very nearly
a) 0 b) 0.7146 c) 0.3 d) 0.2854

4.8 The power requirement in terms of hp is very nearly
a) 350 b) 272.5 c) 300 d) 375

4.9 At 130 psia and 100 ^{0}F, the internal energy in Btu/lb of the refrigerant is very nearly
a) 110 b) 106.94 c) 108.06 d) 112.2

4.10 If the system were to operate as a heat pump, the coefficient of performance will very nearly be
a) 6.4 b) 5.19 c) 6.19 d) 5.5

5: Propionic acid can be produced by the following second order reversible reaction.

$$C_2H_5COONa + HCl \leftrightarrows C_2H_5COOH + NaCl$$

$\Delta H_f(kcal/g\cdot mol)$	- 171.01	- 39.85	-121.7	-97.32

For equimolar concentrations of the reactants and zero concentrations of the products at the start of the reaction, the rate expression in terms of conversion is given by

$$-r_A = k_1 C_{AO}^2\left\{\left(1-X_A\right)^2 - \frac{X_A^2}{K}\right\}$$

where k_1 = forward reaction rate constant, k_2 = rate constant for the reverse reaction, K = equilibrium constant , X_A = conversion and C_{AO}= initial concentrations of the reactants. In figure Kinetics 3 given later are plotted reaction rate data as $\frac{1}{-r_A}$ vs X_A. The initial concentrations of sodium propionate and hydrochloric acid were 2.7 g-mol/L each. The solution of the rate equation in terms of conversion with initial conditions $X_A = 0$ at $t = 0$ and $\frac{dX_A}{dt} = 0$ at equilibrium and with restrictions that $C_{AO} = C_{BO}$, $C_{RO} = C_{SO} = 0$ is given by

$$\ln\frac{X_{Ae} - (2X_{Ae} - 1)X_A}{X_{Ae} - X_A} = 2k_1 C_{AO}\left(\frac{1}{X_{Ae}} - 1\right)t$$

where X_{Ae}= equilibrium conversion = 0.8. A plot of $\ln\frac{X_{Ae} - (2X_{Ae} - 1)X_A}{X_{Ae} - X_A}$ vs t gives a straight line of slope 0.032. Other data are as follows: C_P = 0.97 kcal/(kg.K) . ΔC_P for reactants and products = 0. ρ of solution 1210 kg/m^3 .

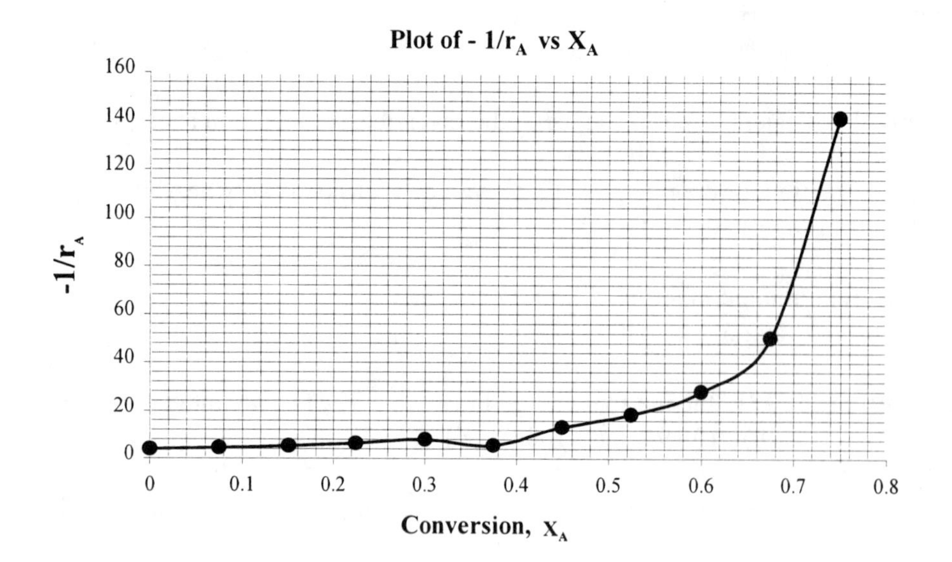

Given these data, select the correct answer for each of the following questions.

5.1 The value of the equilibrium constant K is very nearly
a) 4 b) 16 c) 8 d) 12

5.2 The value of the forward reaction rate constant in units of L/(g-mol.min) is very nearly
a) 0.0237 b) 1.422 c) 0.0474 d) 0.237

5.3 For a 75 % conversion, the value of Damkoheler number is closest to:
a) 3.9 b) 1.95 c) 15.2 d) 0.39

5.4 1480 kg/h of Propionic acid is to be produced in a batch reactor by charging sodium propionate and HCl in equimolar proportion so that the concentration of each reactant will be 3.3 g-mol/L. The feeds contain no products initially. The turnaround time is 40 minutes. The conversion is to be 75 %. The kg of sodium propionate to be charged to the batch reactor will be
a) 2560 b) 3840 c) 2220 d) 2960

5.5 In item 5.4 above, if 20 % free space is allowed, the actual reactor volume in m^3 will be nearly
a) 12 b) 10 c) 15 d) 9

5.6 Propionic acid is to be produced at the same rate, conversion and initial concentrations of reactants and products as in 5.4 above using a CSTR. the working volume of the back-mix continuous reactor in m^3 will be
a) 47.3 b) 40 c) 15 d) 59.1

5.7 Space time in minutes for the CSTR in 5.6 above is nearly
a) 350.7 b) 300 c) 106.3 d) 327.2

5.8 For the same conditions as in 5.6 above, the space-time of a plug flow reactor in minutes will be

a) 29.8 b) 24.6 c) 49.2 d) 14.9

5.9 For the same conditions as in 5.6 above, the volume of plug-flow reactor in m^3 will be

a) 47.3 b) 13.2 c) 6.6 d) 49.2

5.10 The CSTR operation in 5.6 is to be carried out at constant temperature of 50 ^{0}C. The heat to be supplied to the reactor in kcal/h is nearly

a) -166000 b) + 166000 c) - 260248 d) 260248

Solutions

1 :

1.1

Amount of blackash = 3160 lb/h

Ba in BaS in blackash = 0.769 x 3160 x (137.4/169.4) = 1971.0 lb

Ba in BaO in blackash = 0.097 x 3160 x (137.4/153.4) = 274.5 lb

$BaSO_4$ corresponding to above Ba in black ash = (1971 + 274.5) x (233.4/137.4) = 3814.4 lb

Unconverted $BaSO_4$ in blackash = 0.012 x 3160 = 38 lb

Total $BaSO_4$ in barites fed to rotary kiln = 3814.4 + 38 = 3852.4 lb

Therefore percent $BaSO_4$ in feed = (3852.4/4000) x 100 = 96.31 %

Answer is b)

1.2

Need to calculate carbon actually utilized in the reduction of $BaSO_4$.

$BaSO_4$ reduced = 3814.4 lb

Carbon used in reduction = 3814.4 x (12 x 4/233.4) = 788.5 lb

Answer is a)

1.3

BaS in blackash = 0.769 x 3160 = 2430 lb

Then water used for extraction of BaS =(85/15) x 2430 = 13770 lb

(water as calculated above includes fresh water added, recycled water and the condensed steam used in heating.)

Solution in sludge = (1 - 0.65) x 2430 x (100/15) = 5670 lb

Inert solids in blackash = 0.072(3160) = 227.5 lb

(Inert solids in blackash consists of inerts from barites and coal)

Carbon in blackash = 0.05 x 3160 = 158 lb

Water insoluble BaO in blackash = 0.0.097 x 3160 = 306.5 lb

Insoluble $BaSO_4$ in blackash = 0.012(3160) = 38 lb

Total amount of water-insoluble solids in blackash = (227.5 +158 + 306.5 + 38)
= 730 lb

Pounds of solution retained per lb of water-insoluble solids
= 5670/730 = 7.767 lb/ lb solids

Answer is a)

1.4

Ba in thick sludge:

Ba in BaS = (137.4/169.4) x 0.35 x 2430 = 689.9 lb

Ba in BaO = (137.4/153.4) x 90.097) x (3160) = 274.5 lb

Total Ba available for forming chloride = 689.9 + 274.5 = 964.4 lb

Then $BaCl_2$ that can be obtained = 964.4 x (208.4/137.4) = 1462.7 lb

Answer is c)

1.5 $BaS + Na_2CO_3 = BaCO_3 + Na_2S$

$BaCO_3$ produced = (197.4/169.4) x 0.65 x 2430 = 1840.6 lb

Answer is d)

1.6

BaS in overflow clear solution = 0.65(2430) = 1579.5 lb
Then theoretical amount of Na_2CO_3 required = (106/169.4) x 1579.5 = 988.4 lb
With 5 % excess , actual Na_2CO_3 used = 1.05 x 988.4 = 1037.8 lb
Then lb of Na_2CO_3 per lb of BaS actually used = 1037.8/1579.5 = 0.657 lb

Answer is b)

1.7

BaS in sludge = (1 - 0.65)(2430) = 850.5 lb
H_2S evolved = (34/169.4) x 850.5 = 170.7 lb
$(NH_4)_2S$ produced per hour = (68 /34) x 170.7 = 341.4 lb/h
$(NH_4)_2S$ production per year with 8000 h / year on-stream factor
= (8000 x 341.4)/2000
= 1365.6 ton/year

Answer is a)

1.8

BaS in solution = 1579.5 lb
BaS converted = 0.95 x 1579.5 = 1500.5 lb
Na_2CO_3 with 5 % excess = 1037.8 lb from item 6 above.
Water with Na_2CO_3 = (70/30) x 1037.8 = 2421.5 lb
Water in BaS solution = (85/15) x 1579.5 = 8950.5 lb
Wash water = 1000 lb per hour = 1000 lb
Total water = 2421.5 + 8950.5 + 1000 = 12372 lb
Na_2S formed = (78/169.4) x 1500.5 = 690.9 lb
Na_2CO_3 consumed = (106/169.4) x 1500.5 = 938.9 lb
Unconverted Na_2CO_3 = 1037.8 - 938.9 = 98.9 lb
Unconvertd BaS = 1579.5 - 1500.5 = 79 lb
Then total solution = 12372 + 79 + 98.9 + 690.9 = 13240.8 lb
Percent of Na_2S = (690.9/13240.8) x 100 = 5.22 % by weight

Answer is c)

1.9

The boiling solution is alkaline like NaOH solution. Therefore monel is recommended

Answer is c)

1.10

SO_2 produced per hour = 160 lb (Given)
SO_2 oxidized = 0.1 x 160 = 16 lb/h
SO_2 converted to sodium bisulfite = 160 - 16 = 144 lb/h
NaOH used in conversion of SO_2 to bisulfite = (40/64) x 144 = 110 lb/h.
NaOH used in conversion of SO_2 to sulfate = (80/64) x 16 = 20 lb/h
Total caustic used per day in the scrubber = 24 x (110 + 20)/ 2000 = 1.32 ton/d

Answer is b)

2:

Use the triangular diagram provided. (Figure 2)

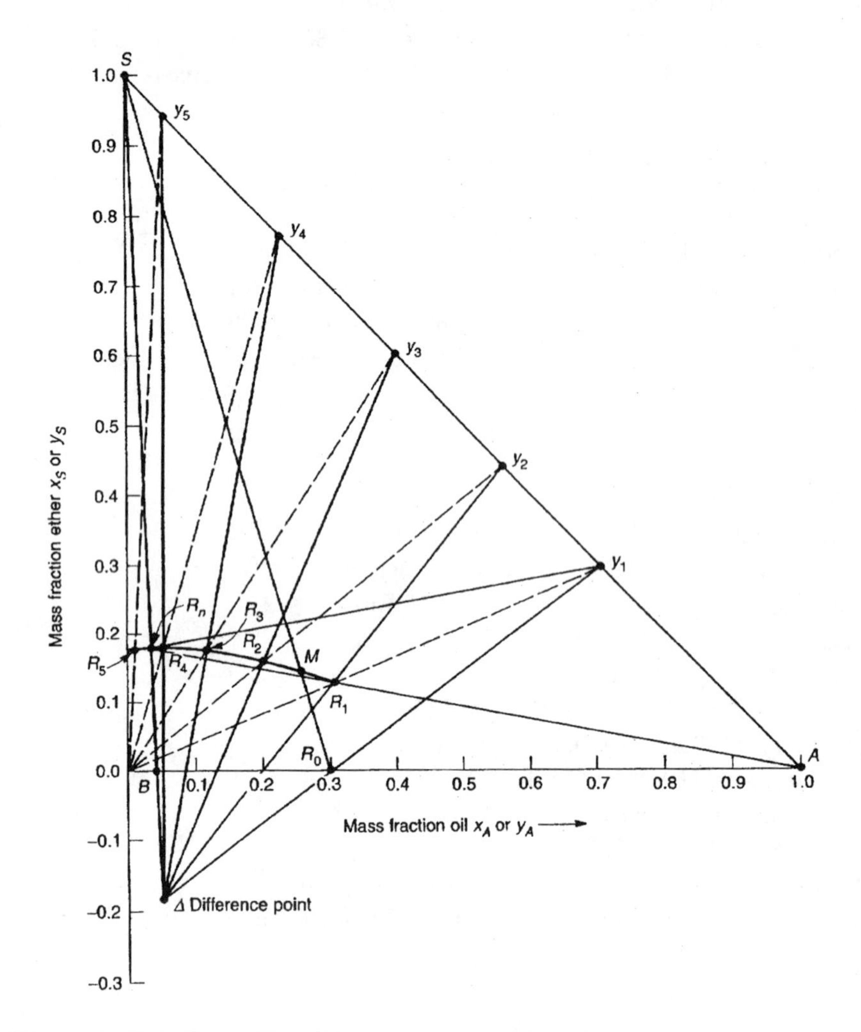

Figure 2: Solution of Problem 2 on triangular diagram

2.1 By material balance , oil in discharged solids = 30 lb/h

Inerts = 0.7(1000 = 700 lb/h

Mass fraction oil = $\frac{30}{700+30}$ = 0.0411 **Answer is d)**

Locate this point on X axis as point B

2.2 Join $\overline{By_{n+1}}$ to intersect underflow curve in R_n

Read composition of R_n as oil = 0.033 ether = 0.179 and inerts = 0.788

solvent in discharged solids = (0.179/0.788)(700) = 159 lb/h **Answer is c)**

2.3 Connect R_n and A and $\overline{y_{n+1}R_0}$ to intersect at M

Minimum solvent-to-solids ratio = $\frac{V_{n+1}}{R_0} = \frac{\overline{(x_A)_0 M}}{\overline{My_{n+1}}} = \frac{1.9}{11.4} = 0.17$

Answeer is a)

2.4 Actual solvent-to-solid ratio = 1.6(0.17) = 0.272

Solvent to be used = 0.272(1000) = 272 lb/h

Answer is c)

2.5 Solvent + solids = 272 + 1000 = 1272 lb/h

Coordinates of the addition point are as follows:

$$\left(x_A\right)_\Delta = \frac{300}{1272} = 0.23 \qquad \left(x_S\right)_\Delta = \frac{272}{1272} = 0.214 \qquad \left(x_I\right)_\Delta = \frac{700}{1272} = 0.55$$

Answer is a)

Locate this point on the diagram as M'.

2.6 Join R_n and M' and extent the line to meet the hypotenuse of the triangle in y_1 i.e. $(y_A)_1$

Also read composition of solution leaving the stage 1 as

$(y_A)_1 = 0.705 \quad (y_S)_1 = 0.295 \quad (y_I)_1 = 0$

mass fraction of ether = 0.295

Answer is d)

2.7 Number of theoretical stages. These constructed on the diagram with the help of underflow locus line and the tie lines. 5 stages are some what better while 4 stages are not good enough.

Number of stages = 4.4 approximately . These are not actual stages. Therefore the number can be fractional

Theoretical stages = 4.4

Answer is c)

2.8 Efficiency is given as 0.75

number of actual stages = 4.4/0.75 = 5.87

actual stages must be a whole number

therefore recommend 6 stages. No additional stage is required.

Answer is a)

2.9 Net flow of the solute towards first stage is given by

$$\left(Dx_A\right)D = \left(V_{n+1}y_A\right)_{n+1} - \left(Rx_A\right)_n = 272 \times 0 - 30 = -3 \qquad \text{lb/h}$$

Answer is b)

2.10 When solvent retained per unit mass of inerts is constant, the locus of the underflow compositions is a straight line given by the equation

$$x_S = \frac{k}{k+1} - x_A$$

Since $x_S = kx_C$ where k = mass of solvent retained per unit mass of inert solids.

This line passes through the points [$x_S = k/(k+1); x_A = 0$], and ($x_S = 0; x_A = 1$).
When mass of solution retained per unit mass of inert solids is constant , the locus of the underflow compositions is parallel to the hypotenuse . Therefore this does not apply to the conditions given in the problem.
In case of nonideal systems the locus of underflow compositions is a curve and not a straight line. Therefore this is ruled out.
In case of ideal systems, tie lines pass through origin "O" representing the inerts (I = 1).

Answer is a)

3:

3.1 Wet bulb absolute saturation = $\frac{43.6}{760-43.6} \times \frac{46}{28} = 0.$ lb ethanol/lb nitrogen

Answer is b)

3.2 $\rho_{nitrogen} = \frac{28}{359}\left(\frac{492}{528}\right) = 0.0727$ lb/ft³

$$\frac{\mu}{\rho D_{AB}} = \frac{0.018 \times 2.42}{0.0727 \times 0.44} = 1.362$$

Answer b)

3.3 Dry bulb absolute saturation

$$\frac{Y-0.1}{122-68} = -\frac{0.37}{367.}$$ from which Y = 0.0443 lb alcohol /lb nitrogen

Answer is a)

3.4 Saturated absolute saturation

$$Y'_S = \frac{229}{760-229} \times \frac{46}{28} = 0.709 \text{ lb alcohol/lb nitrogen.}$$

Answer is b)

3.5 percentage saturation = $\frac{0.0443}{0.709} \times 100 = 6.25$ %

Answer is b)

3.6 Percent relative saturation

$$\frac{p_A}{760-p_A} \times \frac{46}{28} = 0.0443 \quad \text{or} \quad \frac{p_A}{760-p_A} = 0.027 \quad p_A = 19.98 \text{ mmHg}$$

Then percent relative saturation = $\frac{19.98}{229} \times 100 = 8.7$ %

Answer is a)

3.7 Humid volume of gas

$$v_H = 359\left(\frac{1}{28} + \frac{0.0443}{46}\right) \times \frac{52}{49} = 14.13 \text{ ft}^3\text{/lb}$$

Answer is a)

3.8 Humid heat of gas

$$C_S = 0.25 + 0.0443(0.38) = 0.267 \quad \text{Btu/(lb.°F)}$$

Answer is b)

3.9 Lewis number for ethanol-nitrogen mixture

$$\text{Lewis number} = \frac{Sc}{\text{Pr}} = \frac{1.362}{0.71} = 1.9$$

Answer is c)

3.10 Heat needed to raise the temperature to 220 °F

$$Q = \frac{1000 \text{ ft}^3 / \text{min} \times 60 \text{ min/h}}{14.13 \text{ ft}^3/\text{lb}} \times 0.267 \text{ Btu/lb°F} \times (220 - 122) = 111108 \text{ Btu/h}$$

Answer is a)

4:

4.1 1 ton of refrigeration = 200 Btu/min

$$\text{Capacity of the unit} = \frac{506.2 \times 8.33 \times 1.02 \times 0.93 \times (45 - 30)}{200} = 300 \ \textit{to}$$

Answer is d)

4.2 Heat removed in the evaporator $= h_1 - h_4 = h_1 - h_3 = 106.00 - 43.4$
$= 62.6$ Btu/lb

Answer is b)

4.3 Mass flow rate of refrigerant $= \frac{300 \times 200}{62.6} = 958.5$ lb/min

Answer is b)

4.4 The energy balance with gas bounded by compressor walls as system reduces to

$$W_c = h_i - h_o = h_1 - h_2$$

By cross interpolation, we obtain h_2 as follows

h at 100 °F and 128.8 psia = -(117.7 - 117.0) +117.7 = 117.08 Btu/lb
h at 110 °F and 128.8 psia = -(120.2 - 119.5) + 120.2 = 119.58 Btu/lb
$h_2 = (119.08 - 117.08)(4.4/10) + 117.08 = 118.18$ Btu/lb
Work of compression = $(h_1 - h_2) = (106.00 - 11118.18) = -12.06$ Btu/lb
Sign is negative because the work is done on the system.

Answer is c)

4.5 The coefficient of performance of the refrigeration system is

$$\beta = \frac{\textit{Heat removed in the evaporator}}{\textit{work of compression}} = \frac{h_1 - h_4}{h_2 - h_1} = \frac{62.6}{12.06} = 5.19$$

Answer is a)

4.6 Carnot coefficient of performance

$T_1 = 20 + 460 = 480$ °R $\quad T_2 = 95 + 460 = 555$ °R

$$\beta = \frac{480}{555 - 480} = 6.4$$

Answer is b)

4.7 Vapor fraction of mixture after valve
Expansion is isenthalpic Therefore $h_3 = h_4$
$43.4 = x(18.4) + (1-x)(106)$
Solving gives x = 0.7146 which is mass fraction of the liquid
Then vapor mass fraction is 1 - 0.7146 = 0.2854

Answer is d)

4.8 Power requirement of the compressor $= \frac{958.5 \frac{lb}{min} \times 12.06 \frac{Btu}{lb}}{42.427 Btu/hp} = 272.5 \ hp$

Answer is b)

4.9 Internal energy $U = H - PV = 117.00 - \frac{130 \times 144 \times 0.37136}{778} = 108.06 \ Btu/l$

Answer is c)

4.10 When the system operates as heat pump, the coefficient of performance

$$\beta = \frac{118.08 - 43.4}{118.08 - 106.00} = 6.1$$

Answer is c)

5:

5.1 Equilibrium conversion is given as 0.8.

$$K = \frac{X_{AE}^2}{(1 - X_{AE})^2} = \frac{0.8^2}{(1-.8)^2} = 16$$

Answer is b)

5.2 From the solution of the rate equation, slope of the line is

$$slope = 2k_1 C_{AO}\left(\frac{1}{X_{Ae}} - 1\right) = 0.032$$

$$k_1 = \frac{0.32}{2C_{AO}\left(\frac{1}{X_{Ae}} - 1\right)} = \frac{0.32}{2 \times 2.7 \times \left(\frac{1}{0.8} - 1\right)} = 0.0237 \ \frac{L}{g\text{-}mol.\min}$$

Answer is a)

5.3 Damkohler number for second order kinetics is given by $kC_{AO}t$ which is dimensionless and depends only on conversion. At 75 % conversion,

$$kC_{AO}t = \frac{\ln \frac{X_{Ae} - (2X_{Ae} - 1)X_A}{X_{Ae} - X_A}}{2\left(\frac{1}{X_{Ae}} - 1\right)} = \frac{\ln \frac{0.8 - (2 \times 0.8 - 1)0.75}{0.8 - 0.75}}{2\left(\frac{1}{0.8} - 1\right)} = 3.892 \doteq 3.9$$

Answer is a)

5.4 For a conversion of 75 %, the Damkohler number is 3.9 (see 6.3 above)

$$\text{Time of reaction} = \frac{3.9}{kC_{AO}} = \frac{3.9}{0.0237 \times 3.3} \doteq 50 \text{ min.}$$

Turnaround time = 40 min. Total batch time = 50 + 40 = 90 min.

For acid production of 1480 kg/h, acid to be produced per batch = 1480(90/60)
= 2220 kg/batch

Sodium propionate to be charged = 2220 x (96/74) /0.75 = 3840 kg/batch

Answer is b)

5.5 Sodium propionate feed = 3840/96 = 40 kmols

Working volume of the reactor = 40/3.3 = 12.1 m^3

With free board of 20 %, actual volume of the reactor = 12.1/0.8 = 15.1 m^3

Answer is c).

5.6 If CSTR is to be used, propionate feed = $\frac{1480}{74} \times \frac{1}{0.75} = 26.$ kmols/h

$$\text{For a CSTR, } \frac{V}{F_{AO}} = \frac{X_{Af} - X_{Ai}}{(-r_{Af})} = \frac{X_{Af}}{(-r_{Af})} \qquad \text{since } X_{Ai} = 0$$

$$= \frac{X_{Af}}{k_1 C_{AO}^2[(1 - X_A)^2 - X_A^2/16]}$$

$$= \frac{0.75}{0.0237 \times (3.3)^2 \times [(1-0.75)^2 - 0.75^2/16}$$

$$V = \frac{0.75 \times 26.7/60}{0.0237 \times 3.3^2 \times [(1-0.75)^2 - 0.75^2/16]} = 47.3 \text{ m}^3$$

Answer is a)

5.7 Space-time for the reactor = $\tau = \frac{C_{AO} \times X_A}{(-r_{Af})} = \frac{3.3 \times 0.75}{0.0237 \times 3.3^2[(1-0.75)^2 - 0.75^2/16]} = 350.7$ mi

Answer is a)

5.8 Space-time for a plug-flow reactor is given by $\tau = C_{AO} \int_0^{X_A} \frac{dX_A}{-r_A}$

$$= C_{AO} \int_0^{X_A} \frac{dX_A}{0.0237 \times 3.3^2\left[\left(1 - X_A^2\right) - X_A/16\right.}$$

On plot - $1/r_A$ vs X_A, divide the curve in 10 equal intervals with h = 0.075 between $X_A = 0$ and $X_A = 0.75$ and find the area under the curve by Simpson's rule. (Note - It is not necessary to have a plot of $-1/r_A$ vs X_A The values of - $1/r_A$ can be obtained by computation at desired X values. the values were obtained by computation and used to apply Simpson's rule as given below.)

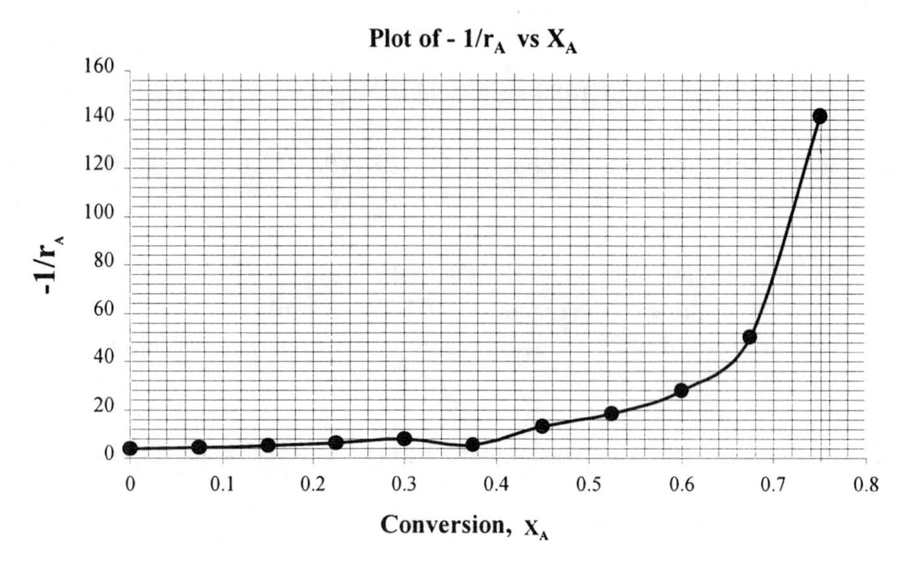

Figure for Problem 5 : Plot of 1/-r_A vs X_A

Area under the curve by Simpson's rule

$$= \frac{0.075}{3}\{3.85 + 2(5.37 + 8.0 + 13.37 + 28.18) + 4(4.53 + 6.49 + 5.46 + 18.59 + 50.22) + 141.7\}$$
$$= 14.9$$

Then $\tau = 3.3 \times 14.9 = 49.2$ min.

Answer is c)

5.9 The volume of a plug-flow reactor is given by

$$V = F_{AO}\int_0^{X_A}\frac{dX_A}{-r_A} = (26.7/60) \times 14.9 = 6.6 \text{ m}^3$$

Answer is c)

5.10 Heat balance on the back-mix reactor gives

$$0 = \rho C_P V_i\left(T_i - T_r\right) - \rho C_P V_o\left(T_0 - T_r\right) + \Sigma Q$$

Let datum temperature be 298 K.
volumetric flow in and out = 26.5/3.3 = 8.03 m³/h
assume feed enters the reactor at 30 °C
ΔH_R = - 121.7 - 97.52 - (- 171.04 - 39.85) = - 8.33 kcal/g-mol
(Heat of mixing ignored.)
mols converted = 26.7x75 = 20 kmol/h = 20000 g-mols/h
ΔH_r = 20000(- 8.33) = - 166000 kcal/h
Substituting in the heat balance equation gives
0 = 1210x8.03x0.97(303 - 298) - 1210x8.03x0.97(323 - 298) + (- 166000) + Q_t
where Q_t = heat transferred to or from the reactor.

Solving for Q_t, $Q_t = +260248$ kcal/h

Plus sign indicates heat is to be supplied to the reactor to maintain the temperature.

Answer is d)

Appendix
Dimensions and Units

Dimensions are names used to describe the characteristics of a physical quantity. Examples of dimensions are length, mass, time, temperature and electric charge. Units are the basic standards of measuring the magnitudes of physical quantities. For example, meter is a unit for the measurement of length.and second is a unit for measuring time. Units are two types:fundamental and derived. Fundamental units are independently defined whereas the derived units are expressed in terms of the fundamental units.Various unit systems differ from one another in the choice of fundamental dimensions and units.(TableI-1).

Table I-1: Dimension and Units

		Metric or SI system:	British	USCS
Dimension:	Symbol	Unit	Unit	Unit
Length	L	Meter	foot	foot
Mass	M	kg	slug*	lb_m
Time	t	Second	Second	Second
Temperature	T	degree K	^{0}F	^{0}F
Force	F *	Newton (N)*	poundal	lb_f

*denotes a derived unit in the system Others are fundamental units

UNITS OF FORCE:

In the SI system, 1 Newton $= (1\ kg) \times (1\ m/s^2)$

In British Engineering system, $1\ slug = \dfrac{1\ pound\text{-}force}{1\ ft/s^2}$

In fps system, Poundal 1 poundal = (1 lb-mass)x($1\ ft/s^2$)

Comparison of British and fps systems shows

1 slug = 32.174 lb_m and 1 lb-force (lb_f) = 32.174 poundals

The US customary system , 1 lb-force (lb_f) = $\dfrac{(1\ lb_m)(1\ ft/s^2)}{g_c}$

where Newton's law proportionality factor $g_c = 32.174\ lb_m.ft/s^2.lb_f$

Molar Units:

A mole of a substance = molecular mass expressed in given units. Molecular mass is also commonly called molecular weight although weight implies force.

1 kilogram-mole = molecular weight in kilograms, 1 pound-mole = molecular weight in pounds.

1 g-mole = molecular weight expressed in grams, 1 kmol = 1000 gmol

For practical purposes, multiples or submultiples of the same unit are used.and are given special names. For example , 1 ft = 12 inches. Here an inch is a unit which has a value of length equal to 1/12 th of one foot.. The metric multiplier factors are given in Table 1-2 and some basic conversion factors are given in Table 1-3.

Table I-2: Multipliers of Standards Units SI or Metric System

Multiplier	prefix of standard unit	symbol	multiplier	prefix of standard unit	symbol
10^9	giga	G	10^{-9}	nano	n
10^6	mega	M	10^{-6}	micro	m
10^3	kilo	k	10^{-3}	milli	m
10^2	hecto	h	10^{-2}	centi	c
10^1	deca	da	10^{-1}	deci	d

Table I-3: some basic conversion factors

Length: 1 meter = 3.281 ft 1 inch = 2.54 cm. = 25.4 mm 1 foot = 12 inches
1 mile = 5280 ft = 1.609 km 1 foot = 30.48
Mass: 1 kg = 2.2046 lb_m 1 lb_m = 453.6 gm = 0.4536 kg 1 U.S. ton = 2000 lb_m
1 metric ton = 1000 kg 1 slug = 32.174 lb_m 1 ton (British) = 2240 lb_m
Time 1 second = (1/3600) h = (1/60) minute
Force: 1 lb_f = 32.174 poundals. = 4.448 N 1 Newton = 10^5 dyn 1 poundal = 0.138 newton
Temperature : degree Kelvin = ^{0}C + 273.16 , ^{0}R = ^{0}F + 459.7
Pressure 1 bar = $1x10^5$ Pa = 1 N/m^2 1 atm. = 1.013 Bar = 1.03 kg/cm^2 1 atm. = 760 mm Hg
Thermal units 1 calorie = 4.1868 J 1 kcal = 4186.8 J 1 Btu = 0.252 kcal = 778.2 ft.lb_f

In practice, one uses conversion factors already tabulated in various references[1,4] Some factors that are required very often are given in Table 1-4.

Table I-4 : Conversion factors

Multiply	By	To obtain	Multiply	By	To obtain
Atm std.	14.70	psi	ft-lb$_f$	0.32	cal
Atm std	760.00	mm Hg	ft-lb$_f$	1.36	joule(J)
Atm std	29.92	in Hg	ft-lb$_f$/s	$1.818x10^{-3}$	hp
Atm std	33.90	ft of water	gal	3.79	liters(L)
Atm std	$1.013x10^{5}$	Pa	hp	33000.00	ft-lb$_f$ /min
Atm std	1.01	Bar	hp	42.40	Btu/min
Atm std	1.03	kg/cm^2	hp	745.70	watt(W)
Atm std	101.30	kPa	hp-h	2545.00	Btu
Bar	$1x10^{5}$	Pa	hp-h	$2.68x10^{6}$	joule(J)
Bar	1.02	kg/cm^2	joule(J)	$9.478x10^{-4}$	Btu
Bar	14.51	psi	joule(J)	0.74	ft-lb$_f$
Btu	1054.80	Joule(j)	joule(J)	1.00	N-m
Btu	$2.928x10^{-4}$	kWh	joule(J)/s	1.00	watt(W)
Btu	778.20	ft.lb$_f$	J/(s-m^2-^{0} C)	1.00	W/m^2-^{0} C
Btu	0.25	kcal	kcal	3.97	Btu
Btu/h	$3.93x10^{4}$	hp	kcal	$1.56x10^{-3}$	hp-h
Btu/h	0.29	watt(W)	kcal	4.19	joule(J)
Btu/h	0.22	ft.lb$_f$ /s	kN/m^2	0.30	in Hg
Btu/h-ft$^{2-0}$ F	4.88	kcal/h-m^2.-^{0} C	kPa	0.15	psi
Btu/h-ft^2 - ^{0} F	$5.67x10^{-3}$	kW/m^2 -^{0} C	kW	1.34	jp
Btu/h-ft^2 - ^{0} F	5.68	J/(s-m^2 -^{0} C)	kW	737.60	ft.lb$_f$ /s
Btu/(h-ft^2 - ^{0} F/ft)	1.49	kcal/h-^{0} C -m	kWh	3413.00	Btu
Btu/(h-ft^2 - ^{0} F/ft)	1.73	J/s-m-^{0} C	kWh	$3.6x10^{6}$	joule(J)
Btu/lb - ^{0} F	1.00	kcal / kg- ^{0} C	kip(K)	1000.00	lb$_f$
cP	$1x10^{-3}$	N-s/m^2	mile	1.61	km
cP	2.42	lb/h-ft	micron	$1x10^{-6}$	meter
cP	0.00	lb/s-ft	poundal	0.14	Newton
cP	3.60	kg/h-m	watts	3.41	Btu/h
cP	1.00	mPa-s	watts	1.00	Joule/s
ft^3	7.48	gal	W/m^2	0.32	Btu/h-ft^2
ft-lb$_f$	$1.285x10^{-3}$	Btu	W/m^2-K	0.18	Btu/h-ft^2 - ^{0} F
ft-lb$_f$	$3.766x10^{-7}$	kWh	W/(m^2-K-/m)	0.58	Btu/(h-ft^2 - ^{0} F/ft)

FE/EIT and Professional Engineering Review Materials

The most comprehensive review books available.

Engineering Press
The oldest and most respected source of FE/PE materials.

Engineer-in-Training License Review

Donald G. Newnan, Ph.D., P.E., Civil Engr., Editor

A Complete Textbook Prepared Specifically For The New Closed-Book Exam

Written By Ten Professors - Each An Expert In His Field

* 220 Example Problems With Detailed Solutions
* 518 Multiple-Choice Problems With Detailed Solutions
* 180-Problem Sample Exam - The 8-hour Exam With Detailed Step-by-Step Solutions
* Printed In Two Colors For Quicker Understanding Of The Content

Here is a new book specifically written for the closed book EIT/FE exam. This is not an old book reworked to look new - this is the real thing! Ten professors combined their efforts so each chapter is written by an expert in the field. These include engineering textbook authors: Lawrence H. Van Vlack *(Elements of Materials Science and Engineering)*, Charles E. Smith *(Statics, Dynamics, More Dynamics)* and Donald G. Newnan *(Engineering Economic Analysis)*.

The Introduction describes the exam, its structure, exam-day strategies, exam scoring, and passing rate statistics. The exam topics covered in subsequent chapters are Mathematics, Statics, Dynamics, Mechanics of Materials, Fluid Mechanics, Thermodynamics, Electrical Circuits, Materials Engineering, Chemistry, Computers, Ethics and Engineering Economy. The topics carefully match the structure of the closed-book exam and the notation of the *Fundamentals of Engineering Reference Handbook.* Each topic is presented along with example problems to illustrate the technical presentation. Multiple-choice practice problems with detailed solutions are at the end of each chapter.

There is a complete eight-hour EIT/FE Sample Exam, with 120 multiple-choice morning problems and 60 multiple-choice general afternoon problems. Following the sample exam, complete step-by-step solutions are given for each of the 180 problems.

60% Text. 40% Problems & Solutions. 8-1/2 x 11, 2-Color,
ISBN 1-57645-001-5 5.0 lbs Yellow/Green **Hardbound** Cover
Order No. 015 784 Pgs 14th Ed 1996 $39.95

Fundamentals of Engineering Video Course

Thousands of engineers have viewed this video series.
Twelve - ½ hour programs prepared for Educational Television Network broadcast
A Video series that excels in graphics.
A golfer hits a ball with an initial velocity of 116 feet per second at an angle of 60 degrees, hoping to clear a grove of trees 10 yards away. The trees are 45 feet high and extend 320 feet in the direction of flight. Will the ball clear the trees? This is one of the many examples presented in the course.

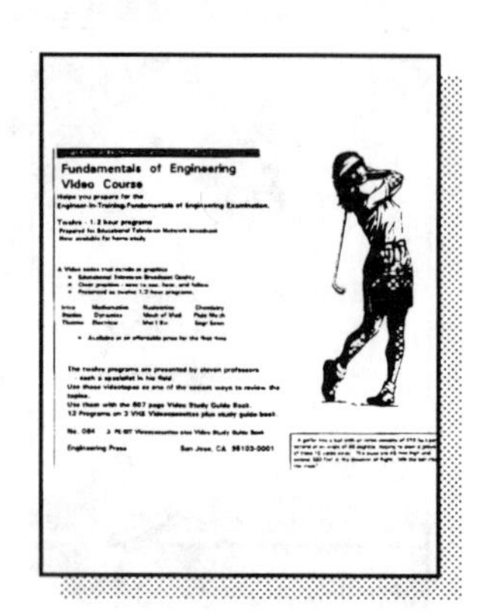

Educational Television Broadcast Quality
Clear graphics - easy to see, hear, and follow
Only the more difficult aspects of each topic are presented, hence a concise review.
Presented as twelve ½-hour programs:

Introduction	Mathematics	Nucleonics	Chemistry
Statics	Dynamics	Mech Mat'l	Fluid Mech
Thermo	Electrical	Mat'l Sci	Engr. Econ

The twelve programs are presented by eleven professors - each a specialist in his field. Students can use these videotapes as one of the easiest ways to review the topics. The 607 page Video Study Guide Book is also included. Use these videos to improve your test performance and your professional skills.

Thousands of engineers have viewed this video series. Now at a new lower price.
ISBN 0-910554-06-4 3 VHS Video Cassettes with Study guide.
Boxed 5.5 lbs 1/CTN
Order No. 064 FE/EIT Video Course $99.00 + $9.50 shipping

Only $99.00. A Great Buy!

Review Materials for the New Discipline Specific Afternoon FE Examination

EIT Civil Review

Donald G. Newnan, P.E.,Civil Engineer, Ph.D., Editor
Written by Six Civil Engineering Professors for the closed book afternoon FE/Civil Examination.
Each topic is briefly reviewed with example problems. Many end-of-chapter problems with complete step-by-step solutions and a complete afternoon Sample Exam with solutions are included.
Soil Mechanics and Foundations
Structural Analysis, frames, trusses
Hydraulics and hydro systems
Structural Design, Concrete, Steel
Environmental Engineering, Waste water
Solid waste treatment
Transportation facilities: Highways, Railways, Airports
Water purification and treatment
Computer and Numerical methods ISBN 1-57645-002-3

Order No. 023 1996 8-1/2 x 11 $29.95

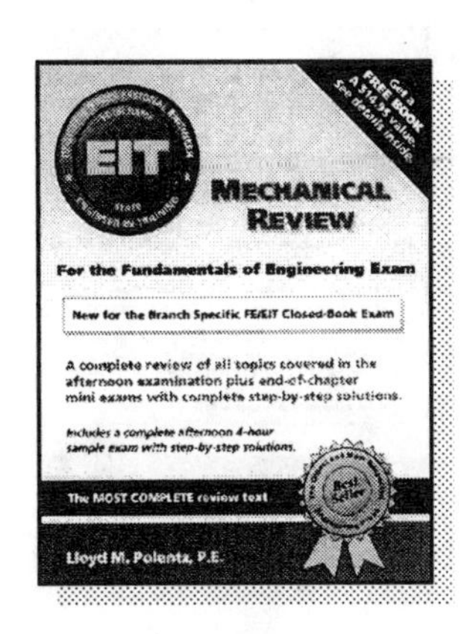

EIT Mechanical Review

Lloyd M. Polentz, P.E.,Mechanical Engineer

Written for the closed book afternoon FE/Mechanical Examination Each topic is briefly reviewed with example problems. Many end-of-chapter problems with solutions and a complete afternoon Sample Exam with step-by-step solutions is included.

Mechanical Design, Dynamic Systems
Vibrations, Kinematics, Thermodynamics
Heat Transfer, Fluid Mechanics
Stress Analysis, Measurement and Instrumentation
Material Behavior and Processing
Computer and Numerical Methods
Refrigeration and HVAC
Fans, Pumps, and Compressors ISBN 1-57645-004-X

Order No. 04X 1996 8-1/2 x 11 $29.95

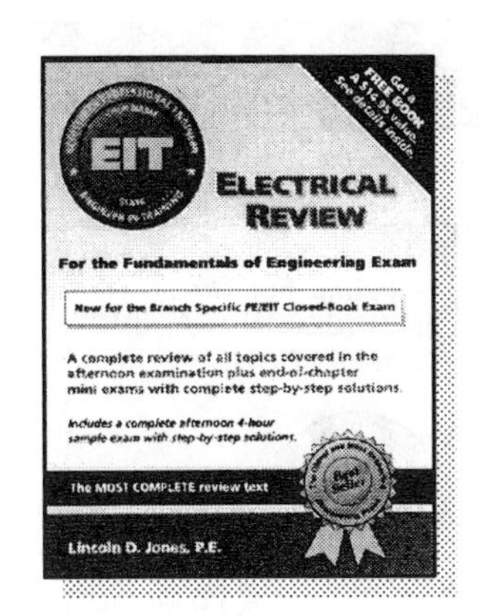

EIT Electrical Review

Lincoln D. Jones, P.E.

Written for the FE/EIT Afternoon Electrical Exam.
Each topic is briefly reviewed with many example problems with complete step-by-step solutions. Each chapter includes end-of-chapter problems with solutions. A complete Sample Exam with solutions is also provided.

The following topics are covered:
Digital Systems, Analog Electronic Circuits
Electromagnetic Theory and Applications, Network Analysis
Control Systems Theory and Analysis, Solid State Electronics and Devices, Communications Theory, Signal Processing
Power Systems, Computer Hardware Engineering
Computer Software Engineering, Instrumentation
Computer and Numerical Methods ISBN 1-57645-006-6

Order No. 066 1996 8-1/2 x 11 $29.95

EIT Industrial Review

Donovan Young, P.E., Ph.D., Industrial Engineer

Written for the FE/EIT Afternoon Industrial Exam.
Each topic is briefly reviewed with many example problems with complete step-by-step solutions. Each chapter includes end-of-chapter problems with solutions. A complete Sample Exam with solutions is also provided.

Production Planning & Scheduling, Engineering Economics
Engineering Statistics, Statistical Quality Control
Manufacturing Processes, Mathematical Optimization & Modeling
Simulation, Facility design and Location, Work Performance & Methods
Manufacturing Systems Design, Industrial ergonomics
Industrial Cost Analysis,Material Handling System Design
Total Quality Management, Computer Computations and Modeling
Queuing Theory and Modeling, Design of Industrial Experiments
Industrial Management, Information System Design
Productivity Measurement and Management ISBN 1-57645-007-4

Order No. 074 1996 8-1/2 x 11 $29.95

EIT Chemical Review

New

D.K Das,P.E.
R.K. Prabhudesai, Ph.D.,P.E.

Written for the FE/EIT Afternoon Chemical Exam.
Each topic is briefly reviewed with many example problems with complete step-by-step solutions. Each chapter includes end-of-chapter problems with solutions. A complete Sample Exam with solutions is also provided.
Material and Energy Balances, Chemical Thermodynamics
Mass Transfer, Chemical Reaction Engineering
Process Design and Economics Evaluation, Heat Transfer
Transport Phenomenon, Process Control
Process Equipment Design, Computer and Numerical Methods
Process Safety, Pollution Prevention ISBN 1-57645-005-8
Order No. 058 1996 8-1/2 x 11 $29.95

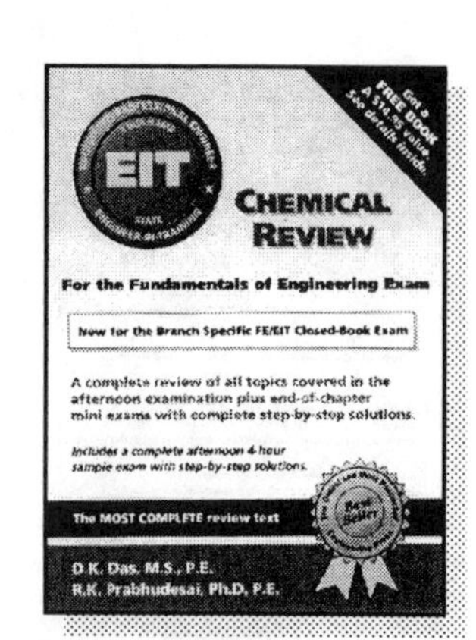

Additional Review Materials for the FE/EIT Examination

Fundamentals Of Engineering Study Guide

Lloyd T. Cheney, P.E.

Written by 10 professors. This is a 2-volume set: 288-page Study Guide plus a 144-page Solution Manual. Review of Math, Chem, Statics, Dynamics, Fluids, Strength of Materials, Thermo, Elect Circuits, Engr. Econ, Computer Sci and Sys plus 550 EIT problems with detailed solutions. 50% Text. 50% Problems and Solutions.
Blue Soft Cover 8-1/2 x 11
2 volume set Shrinkwrapped 3 lbs 12/CTN
ISBN 0-910554-70-9
Order No. 709 432 Pgs $19.50

Preparing For The
ENGINEER-IN-TRAINING EXAMINATION

Irving J. Levinson, P.E.

640 E.I.T. Problems Each With A Detailed Step-By-Step Solution

Preparing For The Engineer-in-Training Exam

Irving Levinson, P.E.

640 Problems And Solutions For the National FE/EIT Exam
Mathematics, Physics, Chemistry, Statics, Dynamics, Mechanics of Materials, Fluid Mechanics, Thermo, Electrical Circuits, and Engineering Economics. 1992
100% Problems and Solutions. 6x9 Paperback 1 lb 22/CTN
Yellow/orange cover **ISBN 0-910554-85-4**
Order No. 854 242 Pgs 3/ed $18.50

Being Successful As An Engineer

W. H. Roadstrum

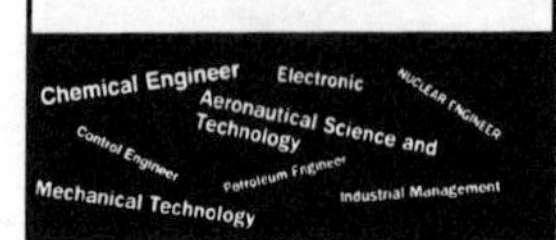

Being Successful As An Engineer

William H. Roadstrum, P.E.

Helps the young engineer make the transition from student to the practice of engineering. Describes engineering and how it functions within an organization.
6 x 9 Paperback 1 lb 36/CTN Blue/white cover
ISBN 0-910554-24-2
Order No. 242 246 Pgs 1988 $14.95

College Reference Series

The Exam File Series are actual midterms and finals from teaching professors around the country. In each exam file you will find excellent review problems with the professor's own solution. Many of these are short and of the type used on the FE/EIT exams. Many students use these in their class studies and when preparing to take the FE exam.

Calculus I Exam File

29 Profs Reveal 301 Exam Problems
With Step-By-Step Solutions. Precalculus, Limits, Continuity, Derivative, and Integrals. **ISBN 0-910554-61-7**
Order No. 617 250 Pages $14.00

Calculus II Exam File

27 Profs Reveal 356 Exam Problems With Step-By-Step Solutions
Transcendental Functions, Integration, Definite Integral, Conics, Polar Coordinates, Indeterminate Forms and Improper Integrals, Taylor Polynomials, and Series. **ISBN 0-910554-62-5**
Order No. 625 282 Pages $15.50

Calculus III Exam File

25 Profs Reveal 333 Exam Problems With Step-By-Step Solutions
Vectors in the Plane and Parametric Equations, Vectors in Space, Functions of Several Variables and Partial Derivatives, and Multiple Integrals.
ISBN 0-910554-63-3
Order No. 633 282 Pages $15.50

Circuit Analysis Exam File

27 Profs Reveal 291 Exam Problems With Step-By-Step Solutions
Circuits, Networks, Transient Analysis, Complex Frequency, 2-Port Networks, State Variables, Fourier, and Laplace. **ISBN 0-910554-53-6**
Order No. 536 314 Pages $15.50

College Algebra Exam File

28 Profs Reveal 509 Exam Problems With Step-By-Step Solutions
Sets, Basic Algebra, Equations & Inequalities, Linear Functions, Polynomial Functions, Linear Systems and Matrices, Analytic Geometry, and Polynomial Eqns. **ISBN 0-910554-77-3**
Order No. 773 410 Pages $16.50

Differential Equations Exam File

27 Profs Reveal 452 Exam Problems With Step-By-Step Solutions.
First Order; Second Order, Solution by Power Series, Laplace Transform, and Applications of Ordinary Differential Eqns. **ISBN 0-910554-70-6**
Order No. 706 506 Pages $18.50

Dynamics Exam File

19 Profs Reveal 333 Exam Problems With Step-By-Step Solutions
Basic Concepts, Kinematics of Particles and Rigid Bodies, Particle Motion Relative to Translating and Rotating Frames, Kinetics of Rigid Bodies, and Vibrations. **ISBN 0-910554-44-7**
Order No. 447 346 Pages $15.50

Engineering Economic Analysis Exam File

16 Profs Reveal 386 Exam Problems With Step-By-Step Solutions
A 32-page review, followed by 386 exam problems and solutions. 30 Interest Tables. Previously published as two separate books: *Engr. Economy Exam File and Engr. Econ Review.* **ISBN 0-910554-83-8 V2**
Order No. 838V2 296 Pages $15.50

Fluid Mechanics Exam File

12 Profs Reveal 203 Exam Problems With Step-By-Step Solutions
Fluid Statics, Motion, Momentum and Energy, Boundary Layer, Conduits, Compressible Flow, and Open Channel Flow.
ISBN 0-910554-48-X
Order No. 48X 218 Pages $13.50

Linear Algebra Exam File

27 Profs Reveal 437 Exam Problems With Step by Step Solutions.
Vectors, Linear Eqns, Matrices, Linear Transforms, Determinates, Eigenvectors, Eigenvalues, and Numerical Methods.
ISBN 0-910554-69-2
Order No. 692 442 Pages $17.50

Materials Science Exam File

21 Profs Reveal 311 Exam Problems With Step-By-Step Solutions
Structure and Behavior of Solids, and Engineering Materials.
ISBN 0-910554-57-9
Order No. 579 314 Pages $15.50

Mechanics of Materials Exam File

16 Profs Reveal 313 Exam Problems With Step-By-Step Solutions
Internal Forces, Stress, Stress-Strain, Axial Loading, Torsion, Bending, Deflection, Combined Stresses, Failure, and Columns.
ISBN 0-910554-46-3
Order No. 463 378 Pages $16.50

Organic Chemistry Exam File

27 Profs Reveal 560 Exam Problems With Step-By-Step Solutions.
ISBN 0-910554-66-8
Order No. 668 534 Pages $20.50

Physics I Exam File: Mechanics

33 Profs Reveal 328 Exam Problems With Step-By-Step Solutions
Vectors, Velocity and Acceleration, Equilibrium, Plane Motion, Work & Energy, Impulse & Momentum, Rotational Dynamics, and Harmonic Motion.
ISBN 0-910554-54-4
Order No. 544 346 Pages $15.50

Physics III Exam File: Electricity And Magnetism

33 Profs Reveal 348 Exam Problems With Step-By-Step Solutions
Coloumb's Law, The Electric Field, The Electric Potential, Capacitance, Charges in Motion, DC circuits and Instruments, Magnetic fields, Magnetic materials, Induced EMF's and Inductance, AC circuits, Electromagnetic Waves and Radiation. **ISBN 0-910554-56-0**
Order No. 560 346 Pages $15.50

Probability And Statistics Exam File

16 Profs Reveal 371 Exam Problems With Step-By-Step Solutions
Statistics, Probability, Discrete & Continuous Distributions, Variances, and More. **ISBN 0-910554-45-5**
Order No. 455 346 Pages $15.50

Statics Exam File

17 Profs Reveal 335 Exam Problems With Step-By-Step Solutions
Force Systems, Equilibrium, Structures, Distributed Forces, Beams, Friction, Virtual Work, and Other Problems.
ISBN 0-910554-47-1
Order No. 471 346 Pages $15.50

Thermodynamics Exam File

12 Profs Reveal 238 Exam Problems With Step-By-Step Solutions
Covers all the topics of a typical first course in thermodynamics.
ISBN 0-910554-49-8
Order No. 498 250 Pages $14.50

Review Materials for the Chemical PE Examination

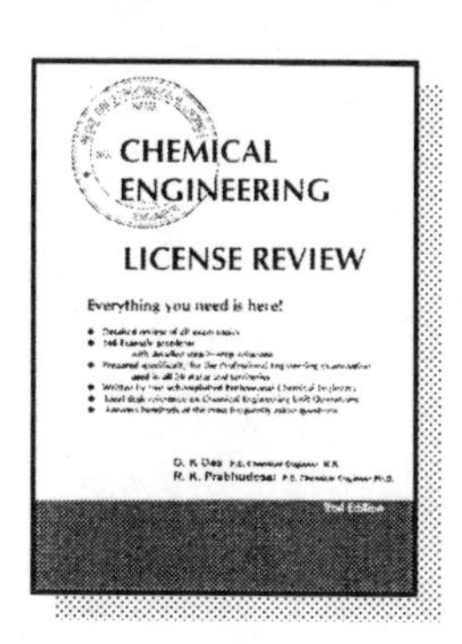

This book was the best seller in its first edition. Now updated with two new chapters. Even better than ever.

Chemical Engineering License Review

D.K. Das P.E.M.S. Chemical Engineer
R.K. Prabhudesai, P.E. Ph.D. Ch.E.☆☆☆☆

Prepared specifically for the PE Exam used in all 50 states. 146 Example problems with detailed step-by-step solutions.
BIG BOOK Format 8-1/2 x 11
Covers ALL topics on the Exam.
Easy-to-use Tables, Charts, and Formulas
Ideal Desk Companion to Perry's *Chemical Engineer's Handbook*
Complete References and Index.17 Chapters:
Units, Fluid Dynamics, Heat Transfer, Evaporation, Distillation, Absorption, Leaching,Liq-Liq Extraction, Psychrometry and Humidification, Drying, Filtration, Thermodynamics,Chemical Kinetics, Process Control, Engineering Economy, Plant Safety, Biochemical Engineering.
The introductory chapter review the test specifications and the author's recommendation on the best strategy for passing the exam. The first chapter review English and SI units and conversions. A complete conversion table is given. The next chapters review fundamentals of fluid mechanics, hydraulics and typical pump and piping problems. Chapter 3 covers heat transfer, conduction, transfer coefficients and heat transfer equipment. Chapter 4 covers evaporation principles, calculations and example problems. Distillation is thoroughly covered in chapter 5. The following chapters cover absorption, leaching, liquid-liquid extraction, and the rest of the exam topics. Each of the topics is reviewed followed by examples of examination type problems. This is the ideal study guide. This book brings all elements of professional problem solving together in one **BIG BOOK**. Ideal desk reference. Answers hundreds of the most frequently asked questions.
The first really practical, no-nonsense review for the tough PE exam. Full Step-by-Step solutions included. ISBN 1-57645-000-7
Order No. 007 1996 2nd Edition 565 Pgs $69.50

NCEES Sample ProblemsAnd Solutions: Chemical Engineering

The Official Publication Of The People Who Write The Exam
State Licensing Requirements; Description of the Exam; Exam Development; Scoring Procedures; Exam Procedures and Instructions; Exam Specifications; Twelve Sample Problems in Chemical Engineering and Solutions; Example Exam Materials.
Order No. 234 172 Pgs 1994 $22.00